Problems in INORGANIC CHEMISTRY

Mrs. Rosenberg

RAJAT PUBLICATIONS
NEW DELHI - 110 002 (INDIA)

RAJAT PUBLICATIONS
4740/23, Ansari Road, Daryaganj
New Delhi- 110 002 (India)
Phones : +91-11-23267924, +91-9811484973
Email: rajat_publication@yahoo.com

Problems in Inorganic Chemistry

Edition 2026

ISBN 81-7880-059-4

PRINTED IN INDIA

Published by Mrs. Seema Wasan for **Rajat Publications**

Preface

This book is an invaluable aid to students in Inorganic Chemistry as a supplement to their textbooks. This includes problems which have been selected from assignments and University examinations. The problems have been arranged in graded manners, i.e., the simple problems have been followed by the difficult problems. All the problems have been solved fully involving simple steps so that the student could be able to solve the similar problems in class as well as in examination. This book is more helpful and instructive than the usual text book solution manuals. This book covers all the assigned topics in the inorganic textbooks. This will help no matter what textbook is used.

Contents

1.	Earlier Concepts of Atoms	1
2.	Modern Concepts of Atomic Structure	17
3.	Atomic Spectroscopy	34
4.	Quantum Chemistry	49
5.	Chemical Bonding	73
6.	Periodic Properties	93
7.	Representative Elements (s–and p-block elements)	107
8.	Transition Metals Including Lanthanides	117
9.	Coordination Compounds and Organometallics	124
10.	Symmetry Properties	129
11.	Solid State Chemistry	149
12.	Nuclear Chemistry	175
13.	Stoichiometry Problems	214
14.	Pollution Problems	227

1
Earlier Concepts of Atoms

Q. 1 : The accompanying figs. show a cathode-ray tube (fig. A) and the deflection of an electron in a region of magnetic field (fig. B). If deflection in, the tube is being produced by the magnetic field alone, predict the effect on the observed deflection by increasing (a) mass of particles, (b) velocity of the particles, (c) magnetic field, and (d) charge on the particles.

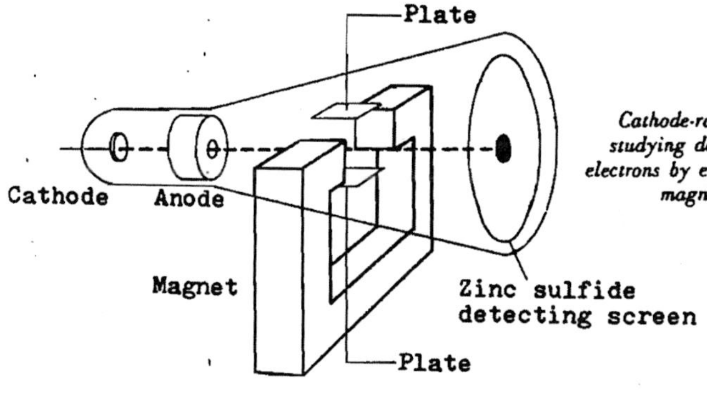

Cathode-ray tube for studying deflection of electrons by electric and magnetic forces.

Fig. A

Solution : The magnetic field, charge, velocity, mass and the deflection of a particle are related by the following equation,

$$Hev = \frac{mv^2}{r}$$

where H is the strength of the magnetic field, e the charge on the particles, r the radius of deflection, v the velocity of the particles, and m the mass of the particles. The derivation of this equation need not be considred in solving this problem.

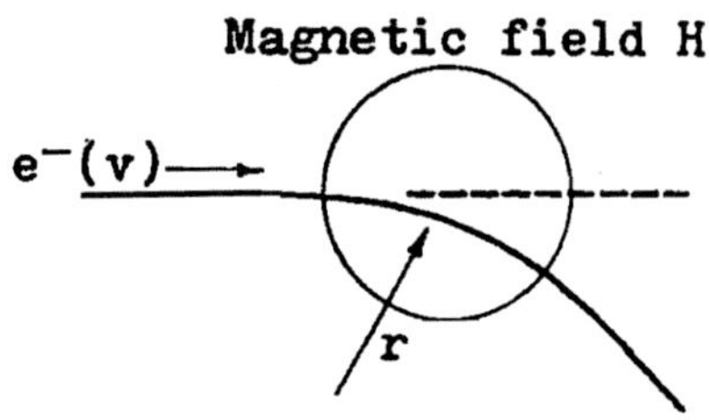

Schematic representation showing how an electron with velocity v is deflected in a region of magnetic field H.

Fig. B

Rewriting the given equation and solving for e/m, the charge to mass ratio,

$$\frac{e}{m} = \frac{v}{Hr}.$$

Since e/m is a constant if (a) the mass is increased, then the radius of deflection must also increase. This is seen more clearly by solving the given equation for r,

$$r = \frac{vm}{He}.$$

Here, if m increases (keeping v, H, and e constnat), then r increases.

Following this type of reasoning, one can predict, (b) the radius of defelction increases with increasing velocity, (c) the radius of deflection decreases with increasing magnetic field strength; and (d) the radius of deflection increases with increasing charge on the particles.

Note that the radius of curvature is inversely proportional to the distance is actually deflected from its original path.

Q. 2 : Consider the experimental apparatus, illustrated in fig. A which is used to measure the deflection of an electron. The source of electrons is a heated filament and the detector consists of a screen painted with a material which phosphoresces when struck by energetic particles. The magnetic field is confined to the spherical region. During a particular run of this experiment, an electron enters the magnetic field at a distance of 30 cm away from the detecting screen and travels 25 cm after leaving the field. If the strength of the magnetic field is 0.18 g/coulomb-sec and the beam deflection on the screen is 4.0 mm. what is the velocity of the electrons in the beam ?

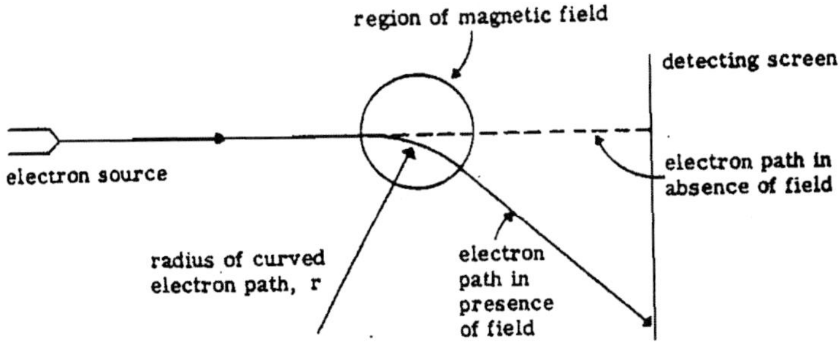

Fig. A

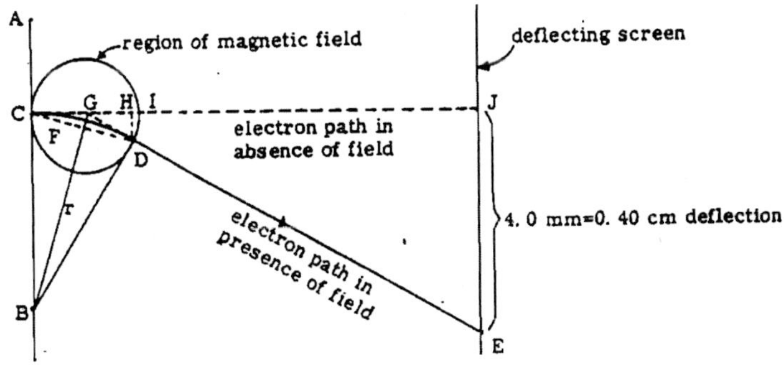

Fig. B

Solution : We are looking for a relationship between the magnetic field strenght H, the deflection of the electron and the electron velocity v. Such a relationship is provided by the equation of motion of an electron in a magnetic field, $v = Her/m$, where e is the electronic charge ($e = 1.602 \times 10^{-19}$ coulomb), r is the radius of curvature, H is the strength of the magnetic field, and m is the electron mass ($m = 9.11 \times 10^{-28}$ g). In order to apply this equation to solve for v, we must first determine r.

In the following discussion to determine r, we will make reference to fig. B (which is not drawn to scale for reasons of clarity). In the problem, we are given that the distance from the point at which the electron beam leaves the field to the deflecting screen is 25 cm. The point at which the electron leaves the field is D. H and D lie on the same vertical line, hence the distance from D to the screen is the same as the distance from H to the screen, or HJ = 25 cm. Since the electron deflection is small, we can assume that the path of the electron is not far from the path the electron would have in the absence of a field. Hence, we can assume that HI is much smaller than CI and HJ, and that FG is much smaller than FB and BG. If we make this assumption that HI is negligible, then $IJ = HJ - HI \cong HJ = 25$ cm. The distance from the point at which the electron enters the field (point C) to the screen is 30 cm, or CJ = 30 cm. Then $CI = CJ - IJ = 30\text{ cm} - 25\text{ cm} = 5$ cm. Denoting the center of the circle containing the magnetic field by G, we have $CG = GI = \frac{1}{2} CI = 5\text{ cm}/2 = 2.5$ cm. Using our assumption on the smallness of HI, we obtain $GH = GH - HI \cong GI = 2.5$ cm. In summary, so far we have obtained the following distances :

$$CG = GI \cong GH = 2.5\text{ cm}, \qquad IJ \cong HJ = 25\text{ cm}.$$

The strategy from this point on will be as follows. We will use the similarity of triangle GJE and triangle GHD to determine HD. Then, using the similarity of triangle CHD and triangle CFB, we will obtain the ratio BF/CF, and from this, the radius of curvature, r.

There is a theorem from elementary plane geometry which states that when a triangle is cut by a line parallel to one of its sides, the two resulting triangles are similar. Since HD and JE are both drawn vertically, and are therefore parallel, triangle GJE is similar to triangle GHD. Hence,

$$\frac{GH}{GJ} = \frac{HD}{JE} \qquad \text{or}$$

$$HD = JE \times \frac{GH}{GJ} = 4 \text{ mm} \times \frac{2.5 \text{ cm}}{27.5 \text{ cm}} = 0.4 \text{ cm} \times \frac{2.5 \text{ cm}}{27.5 \text{ cm}}$$

$$= 0.036 \text{ cm.}$$

Next we note that, using the theorems of plane geometry, it can be shown that triangle CHD is similar to CGB. Then

$$\frac{HD}{CG} = \frac{CD}{BG}$$

Because r = BG, we solve this ratio for BG. We have already determined HD and CG and can find CD by using the Pyhthagorean Theorem, thus, we can find BG. Solving for CD :

$$CD^2 = CH^2 + HD^2 = (5.0 \text{ cm})^2 + (0.036 \text{ cm})^2 = 25.001 \text{ cm}^2;$$

$$CD = 5.00 \text{ cm.}$$

Solving for BG ;

$$BG = \frac{CD.CG}{HD} = \frac{(5.0 \text{ cm})\ (2.5 \text{ cm})}{(0.036 \text{ cm})} = 347.2 \text{ cm.}$$

We can now substitute r into the equation of motion of an electron in a magnetic field. He thus obtain the electron velocity,

$$v = \frac{Her}{m} = \frac{0.18 \text{ g / coulomb} - \text{sec} \times 1.602 \times 10^{-19} \text{ coulomb} \times 347.2 \text{ cm}}{9.11 \times 10^{-28} \text{ g}}$$

$$= 1.10 \times 10^{10} \text{ cm / sec.}$$

Q. 3 : If the gold foil in a Rutherford type experiment is 1/10,000 in. thick, what is a probable minimum number of gold atoms an alpha particle passed through before hitting the fluorescent screen ? 1 in = 2.54 cm, 1 cm = 10^8 Å. The radius of a gold atom is 15 Å.

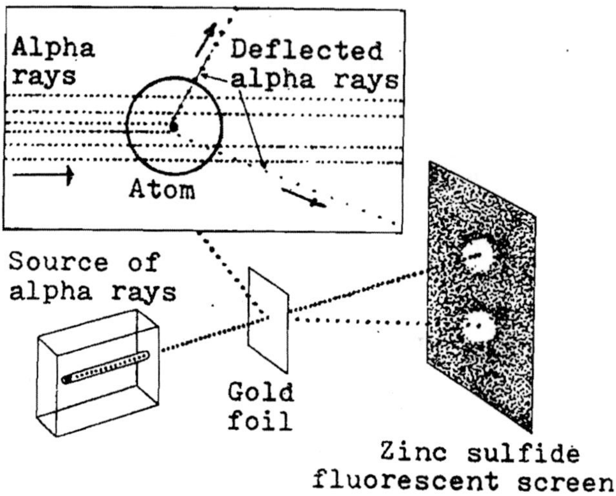

Rutherfored experiment

Solution : In Rutherford's experiment, a beam of alpha particles is aimed at a gold foil of given thickness. Most of the alpha particles pass through undeflected. The deflections that occur suggest that the alpha particles collided with a solid body. This body is the gold nucleus. This experiment elucidated some important properties of the nucleus. One of these properties is the positive charge of the nuclues.

If one knows the thickness of the foil and the atomic radius of the particles that compose the foil, one can calculate the number of particles which make up the width o f the foil. If one assumes cubic packing in the foil, the sum of the lengths of the diameters of the gold atoms will be equal to the thickness of the foil. Proceed as follows :

The radius, and therefore the diameter, of the gold atom is given in terms of angstroms. The thickness of the foil must be converted to the same units.

$$\text{The thickness} = (10^{-4}\ \text{in})\ (2.54\ \text{cm/in})\ (1 \times 10^{8}\ \text{Å/cm})$$
$$= 2.54 \times 10^{4}\ \text{Å}.$$

Diameter of gold nucleus = 2 × Radius = 2 × 1.5 Å = 3 Å.

The minimum number of gold atoms passed by alpha particles is equal to

$$\frac{2.54\times10^{4}\,\text{Å}}{3\text{Å}}=8.466\times10^{3}.$$

Q. 4 : A chemist has a piece of foil that is approximately 5×10^4 atoms thick. If an alpha particle must come within 10^{-12} cm of a nucleus for deflection to occur, what is the probability that an alpha particle will be deflected, assuming the nuclei are not directly behind one another ? Assume that the area of one atom is 4×10^{-16} cm^2.

Solution : To calculate the chance of deflection occurring in any one layer of the foil, assume that each layer is one atom thick. The foil is, therefore, 5×10^4 layers thick. If the number of deflections in any one layer is known, it can be multiplied by 5×10^4 to find the total number of possible deflections as an alpha particle moves through the foil. To find the chance of a deflection occurring in one layer, assume that the atoms are spherical. At 10^{-12} cm from the center of the circle, a deflection can occur. The radius of the circle for which a deflection will occur is 10^{-12} cm. The area of a circle is equal to πr^2 where $\pi = 3.14$ and r = radius.

The area of one atom in which deflection can occur is equal to $3.14 \times (10^{-12})^2$ or 3.14×10^{-24} cm^2. The area of each atom is 4×10^{-16} cm^2. The thickness of each layer is one atom. Therefore, the chance of getting a deflection in any given layer of atoms is

$$\frac{3.14 \times 10^{-24}\ \text{cm}^2}{4 \times 10^{16}\ \text{cm}^2}=7.85 \times 10^{-9}.$$

There are 5×10^4 layers. Thus, a chance of deflection in one of the 5×10^4 layers is

$$5 \times 10^4 \times 7.85 \times 10^{-9} = 4 \times 10^{-4} \text{ or } 1 \text{ in } 2500.$$

Q. 5 : During run of the Millikan oil-drop experiment, the following charges were observed on five different oil droplets : 1.44×10^{-18}, 2.56×10^{-18}, 4.80×10^{-19}, and 9.60×10^{-19} coulomb. Does this support the charge of an electron as being 1.60×10^{-19} coulomb ?

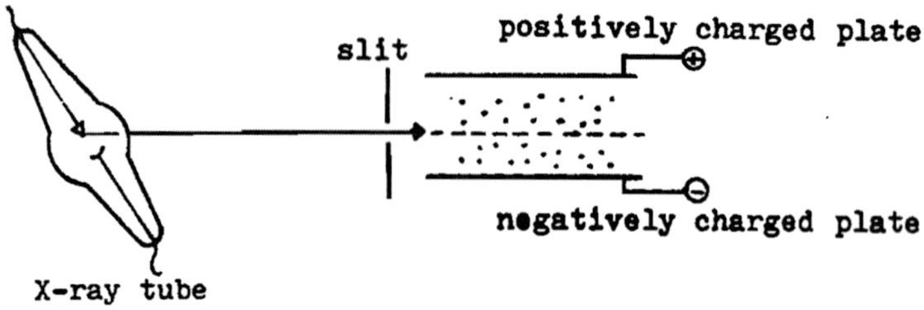

Solution : This experiment, illustrated schematically in the above diagram, was first performed in 1909 by Robert A. Millikan. Oil droplets are sprayed between a top plate having a positive charge and a bottom plate having a negative charge. A beam of X-rays is simultaneously directed at this spray. The X-rays will knock electrons out of molecules in the air and these electrons will be picked up by the oil droplets, which will acquire a charge that is an integral multiple of the electronic charge. This occurs because each oil droplet will pick up an integral number of electrons. By adjusting the charge on the plates, the electons. By adjusting the charge on the plates, the upward pull on the electrons due to attraction by the positive plate on the top and repulsion by the negative plate on the bottom can be balanced against the downward pull of gravity. The forces are balanced when the oil drop is stationary, as viewed through a telescope. By equating the gravitational and the electrostatic forces one can determine the charge on oil droplet.

It is given that the following charges were observed on oil droplets: 1.44×10^{-18}, 2.56×10^{-18}, 4.80×10^{-19}, and 9.60×10^{-19} coulomb. If the elementary electronic charge is 1.60×10^{-19} coulomb, each of these observed charges must be an integral multiple of the elementary charge. Hence, each of these charges must be exactly divisible by 1.60×10^{-19} coulombs.

$$\frac{1.44 \times 10^{-18} \text{ coulomb}}{1.60 \times 10^{-19} \text{ coulomb}} = 9.00$$

$$\frac{2.56 \times 10^{-18} \text{ coulomb}}{1.60 \times 10^{-19} \text{ coulomb}} = 16.00,$$

$$\frac{4.80 \times 10^{-19} \text{ coulomb}}{1.60 \times 10^{-19} \text{ coulomb}} = 3.00. \quad \text{and}$$

$$\frac{9.60 \times 10^{-19} \text{ coulomb}}{1.60 \times 10^{-19} \text{ coulomb}} = 6.00.$$

Each of the observed charges is an integral multiple of 1.60×10^{-19} coulomb, supporting the contention that 1.60×10^{-19} coulomb is the charge of an electron.

Q. 6 : What are the wavelengths of (a) a photon, and (b) an electron, each having kinetic energy of 1 eV ?

Solution : (a) The energy of a photon is given by

$$E = h\upsilon \qquad (1)$$

where h = Planck's constant = 6.6256×10^{-34} Js

ν = frequency

The frequency is defined to be $\frac{c}{\lambda}$, where c = speed of light λ = wavelenght.

Therefore, equation (1) becomes

$$E = h\frac{c}{\lambda}$$

or

$$\lambda = \frac{hc}{E} \qquad (2)$$

Solving equation (2) with E given as

$1 \text{ eV} = 1.6021 \times 10^{-19}$ J yields

$$\lambda = \frac{(6.6256 \times 10^{-34} \text{ Js})\ (2.9979 \times 10^{8} \text{ ms}^{-1})}{1.6021 \times 10^{-19} \text{ J}}$$

$$= 12.398\ 10^{-7} \text{ m}$$

$$\therefore \quad \lambda = 1239.8 \text{ nm}$$

(b) The energy of an electron is defined as

$$E = \frac{1}{2}mv^2 \qquad (3)$$

where m = mass of electron

v = velocity.

Solving equation (3) for v gives

$$v = \left(\frac{2E}{m}\right)^{1/2}$$

$$= \left[\frac{(2)\ (1.6021 \times 10^{-19}\ J}{9.11 \times 10^{-31}\ kg}\right]^{1/2}$$

$$= 5.930 \times 10^{5}\ ms^{-1}$$

The wavelength of an electron is given by $\lambda = \frac{h}{mv}$

$$\lambda = \frac{6.6256 \times 10^{-34}\ Js}{(9.110 \times 10^{-31}\ kg)\ (5.930 \times 10^{5}\ ms^{-1})}$$

$$= 1.226\ 10^{9}\ m$$

$$\therefore \qquad \lambda = 1.226\ nm$$

Q. 7 : From Planck's Law, compute the Stefan-Boltzmann equation for the total density of black-body radiation as a function of temperature. That is,

$$E(T) = \int_0^{\infty} E(\upsilon) dv$$

Solution : If a collection of N oscillators having a fundamental vibration frequency υ takes up energy only in increments of $h\upsilon$, the allowed energies are 0, $h\upsilon$, $2h\upsilon$, $3h\upsilon$, ... Using the Stefan-Boltzmann equation with N_0 = the number of systems in the lowest energy state, the number N_i having an energy ε_i above this ground state is

$$N_i = N_0\ e^{-\varepsilon_i}/kT$$

For a collection of oscillators, the average energy is given as

$$\bar{\varepsilon} = \frac{h\upsilon}{e^{h\upsilon}/kT - 1} \qquad (1)$$

Using equation (1) Planck derived an energy-distribution formula in agreement with the experimental data for black-body radiation. The energy density $E(\upsilon)d\upsilon$ is the number of oscillators per unit volume υ and $\upsilon + d\upsilon$. Planck's law then is given as

$$E(\upsilon)d\upsilon = \frac{8\pi h\upsilon^3}{c^3}\frac{d\upsilon}{e^{h\upsilon/kT}-1}$$

where h = Planck's constant

υ = frequency

c = speed of light

k = Boltzmann's constant

T = absolute temperature

The Stefan-Boltzmann equation for the density of black-body radiation as a function of temperature is then given by

$$E(T) = \int_0^\infty \frac{8\pi h\upsilon^3}{c^3}\frac{d\upsilon}{e^{h\upsilon/kT}-1} = \frac{8\pi h}{c^3}\int_0^\infty \frac{\upsilon^3 d\upsilon}{e^{h\upsilon/kT}-1} \qquad (2)$$

Now, let $y = \frac{h}{kT}$, then $y^3\left(\frac{kT}{h}\right)^3 = \upsilon^3$, and $dy = \frac{h}{kT}d\upsilon$ or $d\upsilon = \frac{kT}{h}dy$. Equation (2) becomes

$$E(T) = \frac{8\pi h}{c^3}\frac{k^3T^3}{h^3}\frac{kT}{h}\int_0^\infty \frac{y^3}{e^y - 1}dy$$

$$= \frac{8\pi k^4T^4}{h^3c^3}\int_0^\infty \frac{y^3}{e^y - 1}dy$$

The solution of the integral is $\frac{\pi^4}{15}$. Therefore,

$$E(T) = \left(\frac{8\pi k^4T^4}{h^3c^3}\right)\left(\frac{\pi^4}{15}\right)$$

$$= \frac{8\pi^5k^4T^4}{15h^3c^3}$$

Q. 8 : Cesium metal is used extensively as the cathode in photocells and in television cameras because it has the lowest inoization energy of all the elements and is readily ionized by light. Given that an incident light of wavelength 5000 Å is incident on a photocell containing cesium, and the threshold wavelength of the photoelectric effect is 6600 Å, calculate the kinetic energy of the ejected electrons from the surface of the cesium.

Solution : The kinetic energy, KE of a photoelectric electron is givne by :

$$KE = h[\nu - (\nu_0)_{Cs}] \qquad (1)$$

where $(\nu_0)_{Cs}$ = the threshold frequency, ν is the frequency before the threshold frequency is reached and h is Planck's constant. But $\nu = \dfrac{c}{\lambda}$, where λ = the wavelength of the incident light and c is the speed of light.

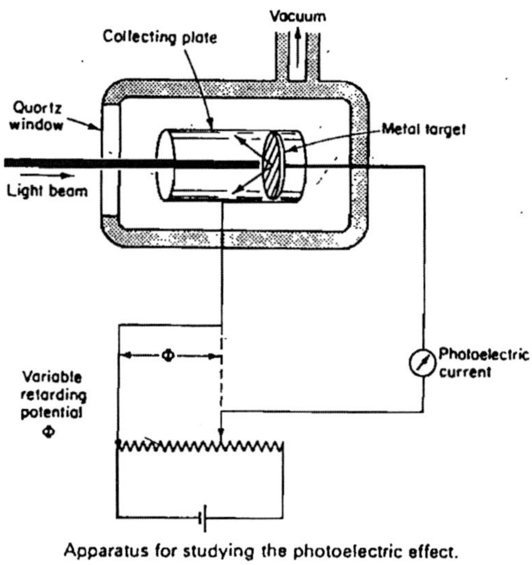

Apparatus for studying the photoelectric effect.

Fig.

$$\therefore \qquad \nu = \frac{\left(3.0 \times 10^{10}\ \frac{cm}{sec}\right)}{(5000\ \text{Å})\left(\dfrac{10^{-8}\ cm}{1\text{Å}}\right)}$$

or $\qquad \nu = 6.0 \times 10^{14}\ sec^{-1}$

$$(\nu_0)_{Cs} = \frac{c}{(\lambda_0)_{Cs}}$$

where $(\lambda_0)_{Cs}$ = the threshold wavelength of cesium

$$\therefore \qquad (\nu_0)_{Cs} = \frac{\left(3.0 \times 10^{10}\ \frac{cm}{sec}\right)}{(6600\ \text{Å})\left(\frac{10^{-8}\ cm}{1\ \text{Å}}\right)}$$

$$\text{or} \qquad (\nu_0)_{Cs} = 4.5 \times 10^{14}\ sec{-1}$$

Using equation (1) gives

$$KE = (6.6 \times 10^{-27}\ erg\ sec)\ [6.0 \times 10^{14}\ sec^{-1} - 4.5 \times 10^{14}\ sec^{-1})$$

$$= 6.6 \times 10^{-27}\ erg\ sec\ (1.5 \times 10^{14}\ sec^{-1})$$

$$\text{or } KE = 9.9 \times 10^{-13}\ erg$$

Q. 9 : Determine the radii of the first and second quantum orbits in the hydrogen atom.

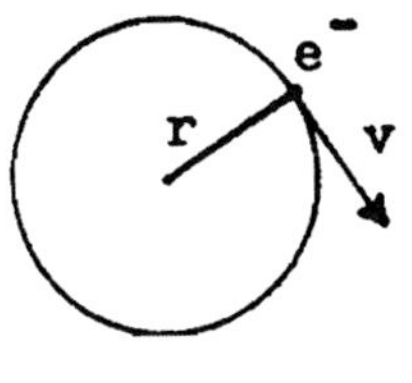

Fig.

Solution : We will derive the Bohr theoretical expression for the radius of the quantum orbit of an electron.

Consider an electron mass m and charge e moving at a constant distance r from the nucleus with a tangential velocity v, as shown in the accompanying figure. In order for this orbit to be stable, the Coulombic force attracting the electron to the nucleus (e^2/r^2) must be balanced by the centrifugal force on the electron (mv^2/r), which, if not compensated for, would compel the electron to leave its orbit. Hence, the magnitude of these two forces must be euqal.

$$\frac{e^2}{r^2} = \frac{mv^2}{r}$$

Solving for r, we obtain $r = \dfrac{e^2}{mv^2}$.

Bohr introduced quantization of the electron's orbit by assuming that the electron's angular momentum (mvr) is given by:

$$mvr = n\left(\frac{h}{2\pi}\right),$$

where h is Planck's constant and n is an index, called the principal quantum number, which takes on the values n = 1 for the first orbit, n = 2 for the second orbit, and so on. Solving for v,

$$v = \frac{nh}{2\pi mr}$$

We can now obtain an expression for r which includes quantization and is in terms of known quantities. Using the above expressions for v and r,

$$r = \frac{e^2}{mv^2} = \frac{e^2}{m}\left(\frac{2\pi mr}{nh}\right)^2 = \frac{4\pi^2 e^2 mr^2}{n^2h^2},$$

$$r = \frac{n^2h^2}{4\pi^2e^2m}$$

The radius of the first orbit (n = 1) is called the Bohr radius. Solving for r : e = 4.80×10^{-10} esu and m = 9.11×10^{-28} g.

$$r_1 = \frac{1^2 \times h}{4\pi^2e^2m} = \frac{1 \times (6.626 \times 10^{-27}\ \text{erg}-\text{sec})^2}{4\pi^2 \times (4.80 \times 10^{-10}\ \text{esu})^2 \times 9.11 \times 10^{-28}\ \text{g}}$$

$$= 5.29 \times 10^{-9}\ \text{cm} = 0.529\ \text{Å}.$$

Let the radius of the nth orbit be denoted r_n. Then

$$r_n = \frac{n^2h^2}{4\pi^2e^2m} = n^2\left[\frac{h^2}{4\pi^2e^2m}\right] = n^2r_1 = n^2 \times 0.529\ \text{Å}.$$

Hence, the radius of the second quantum orbit is

$$r^2 = n^2 \times 0.529\ \text{Å} = (2)^2 \times 0.529\ \text{Å}$$

$$= 2.12\ \text{Å}.$$

Q. 10 : At what value of Z would the innermost orbit be expected to be pulled inside a nucleus of radius 1.0×10^{-13} cm, assuming simple Bohr theory and Bohr radius = 0.5292×10^{-8} cm ? (Assume N = 1).

Solution : Niels Bohr stated, for a hydrogen atom, that only certain permitted values for the radius of the electron path exist. He found that the radius, r, of this electron path was equal to a whole number, N, squared, divided by the atomic number, Z, times the Bohr radius, a_0, which is constant. That is

$$r = \frac{N^2}{Z} a_0.$$

The above question, can be answered by using this equation. N is equal to the quantum number corresponding to the energy level occupied by the electron. It has values of n = 1, 2, 3, 4, ... One is given r and a_0 and is asked to compute Z. The electron occupying the innermost orbital is in the ground state, thus N = 1. Thus,

$$Z = \frac{N^2}{r} a_0 = \frac{(1)^2}{1 \times 10^{-13} \text{ cm}} \times .5292 \times 10^{-8} \text{ cm}$$
$$= 5.292 \times 10^4.$$

Q. 11 : Assuming the Bohr radius is 0.5292×10^{-8} cm, what frequency of a light wave has the same wavelength as the Bohr radius ?

Solution : The question asks one to determine the frequency of a light wave with a wavelength equal to the Bohr radius, $.5292 \times 10^{-8}$ cm. In answering this question, it should be remembered that there is a relation between frequency and wavelength, $c = \lambda\nu$, where c = the speed of light (2.998×10^{10} cm/sec), λ = wavelength, and ν = frequency. One can solve ν in this equation by substituting in the values of c and λ. Thus

$$\nu = \frac{c}{\lambda} = \frac{2.998 \times 10^{10} \text{ cm/sec}}{.5292 \times 10^{-8} \text{ cm}} = 5.665 \times 10^{18} \text{ sec}^{-1}.$$

Q. 12 : The atomic radius of hydrogen is 0.037 nm. Compare this fig. with the length of the first Bohr radius. Explain any differences.

Solution : The first Bohr radius, r, is equal to $\frac{N^2}{Z} a_0$, where a_0 is .055292 nm, N is an integer, and Z the atomic number. For hydrogen, Z = 1. For the first Bohr radius N = 1. Thus

$$r = \frac{(1)^2}{1} a_0 = .05292 \text{ nm}.$$

Thus, the atomic radius of hydrogen (.037 nm) is considerably smaller than the first Bohr radius. This can be explained by the fact that the atomic radius is calculated from the bond length of a H_2 molecule. There is considerable orbital overlap in bonded atoms, which means the bond length is shorter than the sum of the radii of lone atoms.

2

Modern Concepts of Atomic Structure

Q. 1 : Alpha particles emitted from radioactive radium has an energy of 4.8 MeV (million electron volts). Calculate the de Broglie wavelength of one of these alpha particles, given the following data :

(mass of alpha particle = 6.6×10^{-24} g

$h = 6.6 \times 10^{-27}$ erg sec

$1.0 \text{ MeV} = 1.6 \times 10^{-6}$ erg

$1 \text{ erg} = 1 \text{ g cm}^2 \text{ sec}^{-2}$).

Solution : Radium is a radioactive element that emits alpha particles (helium nuclei, He^{2+}), beta particles (elctrons) and gamma rays. Energy depends on frequency or wavelength according to the relation :

$$E = h\nu = \frac{hc}{\lambda} \qquad (1)$$

From Einstein's formula :

$$E = mc^2 \qquad (2)$$

where m = the mass of the particle and c is the speed of light; the particle's momentum is p = mc.

$$\therefore \qquad p = mc = \frac{E}{c} \qquad (3)$$

Substituting (1) into equation (3) yields

$$p = \frac{h}{\lambda}$$

but $\frac{h}{\lambda}$ is also equal to mv hence

$$p = \frac{h}{\lambda} = mv.$$

Equation (4) is the de Broglie equation. Rearranging equation (4) yields

$$\lambda = \frac{h}{mv} \qquad (5)$$

The kinetic energy, KE, of the alpha particles is given as 4.8 MeV. That is,

$$KE = 4.8 \text{ MeV}.$$

But, $KE = \frac{1}{2}mv^2$

$\therefore$ $\frac{1}{2}mv^2 = 4.8 \text{ MeV}$

and $v^2 = \frac{2(4.8 \text{ MeV})}{m}$

$$= \frac{(9.6 \text{ MeV})\left(\frac{1.6\times10^{-6} \text{ erg}}{1 \text{ MeV}}\right)\left(\frac{1 \text{ g cm}^2 \text{ sec}^{-2}}{1 \text{ erg}}\right)}{6.6 \times 10^{-24} \text{ g}}$$

$$= 2.32 \times 10^{18} \frac{\text{cm}^2}{\text{sec}^2}$$

and $v = \sqrt{2.32\times10^{18} \frac{\text{cm}^2}{\text{sec}^2}}$

$$v = 1.52\times10^9 \frac{\text{cm}}{\text{sec}}$$

This is the velocity of the alpha particles. Substituting this velocity into equation (5), we have

$$\lambda = \frac{(6.6 \times 10^{-27} \text{ erg sec})\left(\frac{1 \text{ g cm}^2 \text{ sec}^{-2}}{1 \text{ erg}}\right)}{(6.6 \times 10^{-24} \text{ g})\left(1.52 \times 10^9 \frac{\text{cm}}{\text{sec}}\right)}$$

or $\lambda = 6.6\times10^{-13}$ cm·

Q. 2 : It has been suggested that the time required for light to travel 1 cm be used as a unit of time called the "jiffy". How many seconds correspond to one jiffy ? The speed of light is about 3.0×10^8 m/sec.

Solution : The relationship between distance, speed, and time is distance = speed × time, or, $\text{time}=\dfrac{\text{distance}}{\text{speed}}$.

The time it takes light to travel 1 cm is :

$$\text{time} = \frac{\text{distance}}{\text{speed}} = \frac{1 \text{ cm}}{3 \times 10^8 \text{ m/sec}}$$

$$= \frac{1 \text{ cm}}{3 \times 10^8 \text{ m/sec} \times 10^2 \text{ cm/m}} = \frac{1 \text{ cm}}{3 \times 10^{10} \text{ cm/sec}}$$

$$= 3.3 \times 10^{-11} \text{ sec.}$$

Hence, 1 jiffy = 3.3×10^{-11} sec.

Q. 3 : Consider a molecule of hydrogen at 478.15 K and moving with a velocity of 2×10^5 cm/sec. What is the de Broglie wavelength of this hydrogen atom ?

Solution : The de Broglie wavelength (λ) is related to the momentum of a particle by the de Broglie equation :

$$\lambda = h/mv,$$

where h is Planck's constant, m is the mass of the particle, and v is its velocity (momentum = mv). Calculate the de Broglie wavelength of the hydrogen atom by directly substituting into this equation.

The mass of a hydrogen atom is $m = 3.34 \times 10^{-24}$ g. Planck's constant is $h = 6.626 \times 10^{-27}$ erg-sec $= 6.626 \times 10^{-27}$ g-cm^2/sec. Hence, the de Broglie wavelength is :

$$\lambda = \frac{h}{mv} = \frac{6.626 \times 10^{-27} \text{ g-cm}^2\text{/sec}}{3.34 \times 10^{-24} \text{ g} \times 2 \times 10^5 \text{ cm/sec}} = 9.9 \times 10^{-9} \text{ cm}$$

$$= 9.9 \times 10^{-9} \text{ cm} \times 10^8 \text{ Å/cm} = 0.99 \text{ Å.}$$

Q. 4 : Determine the de Broglie wavelength of an electron (mass $= 9.11 \times 10^{-28}$ g) having a kinetic energy of 100 eV.

Solution : Using the value for the kinetic energy of the electron, we will calculate its momentum and substitute this into the de Broglie equation to get the wavelength of the corresponding matter wave.

The kinetic energy, k, of a particle of mass, m, moving with velocity v is $k = ½ mv^2$ and its momentum, p, is $p = mv$. Kinetic energy and momentum are related as follows :

$$k = ½mv^2 = \frac{1}{2}m^2v^2 = \frac{1}{2m}(mv)^2 = \frac{p^2}{2m} = k \text{ or } p = (2mk)^{½}.$$

The kinetic energy of our electron is k = 100 eV = 100 eV × 1.602×10^{-12} erg/eV = 1.602×10^{-10} erg. Hence, its momentum is :

$$p = (2mk)^{½} = (2 \times 9.11 \times 10^{-28} \text{ g} \times 1.602 \times 10^{-10} \text{ erg})^{½}$$

$$= (.2919 \times 10^{-36})^{½} \text{ erg}^{½} \text{ g}^{½}$$

$$= (.2919 \times 10^{-36})^{½} (\text{g} - \text{cm}^2/\text{sec}^2)^{½} \text{ g}^{½}$$

$$= 5.403 \times 10\text{–}19 \text{ g} - \text{cm/sec}.$$

The de Broglie equation is :

$$\lambda = h/mv = h/p,$$

where h is Planck's constant ($h = 6.626 \times 10^{-27}$ erg/sec) $= 6.626 \times 10^{-27}$ g – cm²/sec. Hence, the de Broglie wavelength of a 100 eV electron is :

$$\lambda = \frac{h}{p} = \frac{6.626 \times 10^{-27} \text{ g} - \text{cm}^2/\text{sec}}{5.403 \times 10^{-19} \text{ g} - \text{cm/sec}} = 1.226 \times 10^{-8} \text{ cm}$$

$$= 1.226 \times 10^{-8} \text{ cm} \times 10^{8} \text{ Å/cm} = 1.226 \text{ Å}.$$

Q. 5 : Calculate the uncertainty in the location of a 100 g turtle crossing a road at a velocity of 1 mm/sec, the uncertainty of the latter being 10^{-4} mm/sec. ($h = 6.626 \times 10^{-27}$ erg–sec and 10 mm = 1 cm).

Solution : This problem demands the use of the Heisenberg Uncertainty Principle. This principle states that one can not simultaneously locate both the position of a particle and its momentum with absolute precision. Whenever one quantity, either position or momentum, is known precisely, the other is known less precisely. Heisenberg showed that the uncertainty in the determination of momentum of a particle, Δp, and the simultaneous determiantion of its location, Δx is related by the equation $\Delta p \Delta x \geq h$, where h is Planck's constant. Momentum p is equal to mv, where m is mass and v the velocity of the mass, thus $\Delta p = m\Delta v$. One substitutes this into the Heisenberg inequality attaining $m\Delta v\Delta x \geq h$. One can now solve for Δx. 1 erg = g cm²/sec², 1 mm = .1 cm.

$$\Delta x = \frac{h}{m\Delta h} = \frac{6.626 \times 10^{-27}\ \text{erg} - \text{sec}}{(100\ \text{g})\ (10^{-44}\ \text{mm/sec})} \cdot \frac{1\ \text{mm}}{.1\ \text{cm}}$$

$$= 6.626 \times 10^{-24}\ \text{cm}.$$

Q. 6 : Consider a kinetic experiment in which the chemical reaction is accompanied by a temperature change. Using the Heisenberg uncertainty principle, calculate the uncertainty in a simultaneous measurement of time and temperature.

Solution : The Heisenberg uncertainty principle relates the uncertainty in energy, ΔE, and the uncertainty in time, Δt, arising from a simultaneous measurement of the two, by the inequality $\Delta EDT \geq h$, where h is Planck's constant. Thus, the product of the two uncertainties cannot be less than Planck's constant. In order to apply the Heisenberg uncertainty principle to our problem, we must relate temperature and energy.

The internal energy of a substance is approximately proportional to its absolute temperature, T. In fact, for ideal gases, the two are exactly proportional. Let the constant of proportionality for our particular system be c. Then, $E = cT$ and $\Delta E = c\Delta T$, where ΔT is the uncertainty in the temperature measurement. Then the uncertainty relation becomes :

$$\Delta E\Delta T = c\Delta T\ \Delta T \geq h \qquad \text{or,} \qquad \Delta T\ \Delta t \geq h/c.$$

Q. 7 : An electron was accelerated through a potential difference of 1.00 ± 0.01 kV. What is the uncertainty of the psoition of the electron along its line of flight ?

Solution: If the momentum of a particle is precisely specified, the position is totally uncertain. This remarkable destruction of one of the fundamentals of classical mechanics is a special case of the Heisenberg Uncertainty Principle. It states that if the momentum is known to lie within a range δp, then the position must be uncertain to an extent δz, where

$$\delta p \delta z \geq \frac{\hbar}{2} \qquad (1)$$

Here $\hbar = \frac{h}{2\pi}$ and h = Planck's constant. Note that z is the line of flight. The uncertainty of momentum can be calculated from

$$\delta p = m_e \delta v \qquad (2)$$

where δv = the uncertainty in velocity. This δv can be calculated from $\frac{1}{2} m_e v^2 = eV$.

For a potential difference of 0.99 kV,

$$v = \sqrt{\frac{2ev}{m_e}} = \sqrt{\frac{2(1.602\times10^{-19}\ C)\ (900V)}{9.110\times10^{-31}kg}}$$

$$= \sqrt{\frac{2(1.602\times10^{-19})\ (990)CV}{9.110\times10^{-31}kg}}$$

$$= \sqrt{3.482\times10^{14}\frac{JV^{-1}V}{kg}}$$

$$= \sqrt{3.482\times10^{14}\frac{kg\ m^2\ sec^{-2}}{kg}}$$

$$= \sqrt{3.482\times10^{14}\ m^2\ s^{-2}}$$

∴ $v = 1.86 \times 10^7\ m\ s^{-1}$

Similarly, for a potential difference of 1.01 kV,

$$v = 1.885 \times 10^7\ m\ s^{-1}$$

Hence $\delta v = (1.885 - 1.866) \times 10^7\ ms^{-1}$

$$= 0.019 \times 10^7\ m\ s^{-1}$$

$$= 1.9 \times 10^5\ ms^{-1}$$

Solving for δp in equation (2) gives

$$\delta p = (9.110 \times 10^{-31}\ kg)\ (1.9 \times 10^5\ m\ s^{-1})$$

$$= 17.309 \times 10^{-26}\ kg\ m\ s^{-1}$$

$= 1.7 \times 10^{-25}$ kg m s^{-1}

Rearranging equation (1) to solve for δz gives

$$\delta z \geq \left(\frac{\hbar}{2}\right) \Big/ \delta p = \frac{1.054\times10^{-34}\,\text{Js}}{2(1.7\times10^{-25}\,\text{kg m s}^{-1})}$$

$$= 3.09\times10^{-10}\,\text{Js / kg m s}^{-1}$$

$$= 3.09\times10^{-10}(\text{kg m}^2/\text{s}^2)\text{s / kg m s}^{-1}$$

or $\delta z = 3.09 \times 10^{-10}$ m = 309 pm = 3.09 Å.

Q. 8 : Apply the Heisenberg Uncertainty Principle to estimate the kinetic energy of the electron in a hydrogen atom at a distance r from the nucleus. Then calculate the equilibrium distance r_e by minimizing the total enery E, kinetic + potential, with respect to r. Compare E calculated in this way with the experimental value of –13.59 eV.

Solution : The Heisenberg Uncertainty Principle specifies that, in simultaneous measurement of the position and momentum of a particle,

$$\Delta r \Delta p \geq \frac{h}{2\pi} \tag{1}$$

where Δr = the uncertainty in position
Δp = the uncertainty in momentum
h = Planck's constant.

Note that the right side of equation (1) can be h, $\frac{h}{2\pi}$ or $\frac{h}{4\pi}$, depending on the precise definition of Δr and Δp. For the H atom, equation (1) will be adequate.

Rearranging equation (1) to solve for Δp gives

$$\Delta p \geq \frac{h}{\Delta r 2\pi} \tag{2}$$

Recall that

$$E = \frac{1}{2}mv^2 = mv(\frac{1}{2}v), \tag{3}$$

where v = velocity

and m = mass.

But $p = mv$ = momentum.

Substitute for p in equation (3) gives

$$E = p\left(\frac{1}{2}v\right) = \frac{p^2}{2m} \quad (4)$$

where $p^2 = m^2v^2$

From equation (4),

$$\Delta E = \frac{(\Delta p)^2}{2m}$$

or $(\Delta p)^2 = \Delta E2m$ (5)

Combining equations (2) and (5) yields

$$(\Delta p)^2 = \Delta E2m \geq \left(\frac{h}{\Delta r2m}\right)^2$$

or $$2m\Delta E(\Delta r)^2 \geq \frac{h^2}{4\pi^2} \quad (6)$$

Rearranging equation (6) to solve for ΔE yields

$$\Delta E \geq \frac{h^2}{8m(\Delta r)^2\pi^2}$$

Therefore, the minimal energy is represented by

$$E_k = \frac{h^2}{8mr^2\pi^2}$$

Total energy is the summation of the kinetic and potential energies. That is :

$$E_T = E_k + E_p$$

$$= \frac{h^2}{8mr^2\pi^2} - \frac{Ze^2}{r}$$

Recall that Z = 1 for an H. Therefore,

$$E_T = \frac{h^2}{8mr^2\pi^2} - \frac{e^2}{r} \quad (7)$$

To find what r_e is, take the partial derivative of equation (7) with respect to r only and set it equal to zero. That is :

$$\frac{\partial E}{\partial r} = -\frac{2h^2}{8m\pi^2 r_e^3} + \frac{e^2}{r_e^2} = 0.$$

Therefore, $r_e = \dfrac{h^2}{4\pi^2 me^2}$

Using this r value in equation (7) gives

$$E = \frac{h^2}{8m\pi^2}\left(\frac{4\pi^2 me^2}{h^2}\right)^2 - e^2\left(\frac{4\pi^2 me^2}{h^2}\right)$$

$$= -\frac{2\pi^2 me^4}{h^2}$$

Here $m = 9.11 \times 10^{-28}$ g

$e = 4.80 \times 10^{-10}$ esu

and $h = 6.62 \times 10^{-27}$ erg sec. Therefore,

$$E = \frac{-2\pi^2(9.11 \times 10^{-28} g)\ (4.80 \times 10^{-10}\ esu)^4}{(6.62 \times 10^{-27}\ erg\ sec)^2}$$

$$= \frac{-2\pi^2(9.11 \times 10^{-28})\ (4.80 \times 10^{-10})^4 g\ esu^4}{(6.62 \times 10^{-27})^2 (erg\ sec)^2}$$

But 1 erg = 1g cm^2 sec^{-2}, and 1 esu^2=1 dyne cm^2=1 g cm sec^{-2} = g cm^3 sec^{-2}. Therefore,

$$E = -2.181 \times 10^{-11} \frac{g\ g^2\ cm^6\ sec^{-4}}{g^2\ cm^4\ sec^{-4}\ sec^2}$$

$$= -2.181 \times 10^{-11}\ g\ cm^2\ sec^{-2}$$

$$= -2.181 \times 10^{-11}\ erg$$

or $E = -13.6$ eV.

Observe that this is very close to the experimental value.

Q. 9 : Discuss the following statement : The Pauli exclusion principle is the main reason why atoms do not collapse to a point.

Solution : The Pauli exclusion principle states that no two electrons in the same atom can be completely identical; that is, have the same values for all four quantum numbers. The quantum numbers measure the energy level of the electron, the order of increasing distance of the average electron distribution from the nucleus, the angular shape of electron distribution, (the electronic magnetism), and the two possible orientations of electron spin. An electron in an atom is completely described by its four quantum numbers. Inherent in this description is the point of location of an electron at a specific time. If two electrons have the same four quantum numbers, they would have the same point of location at a specific time. This means, therefore, that all the electrons would collapse to a point at a specific time. The Pauli exclusion principle states that no two electrons can have the same four quantum numbers and therefore the atom cannot collapse to a point.

Q. 10 : Apply Hund's rules to obtain the electron configuration for Si, P, S, Cl, and Ar.

Si	$3p^2$	↑	↑	___
P	$3p^3$	↑	↑	↑
S	$3p^4$	↑↓	↑	↑
Cl	$3p^5$	↑↓	↑↓	↑
Ar	$3p^6$	↑↓	↑↓	↑↓

Solution : The ground state of an atom is that in which the electrons are in the lowest possible energy level. Each level may contain two electrons of opposite spin. When there are several equivalent orbitals of the same energy, Hund's rules are used to decide how the electrons are to be distributed between the orbitals :

(1) If the number of electrons is equal to or less than the number of equivalent orbitals, then the electrons are assigned to different orbitals.

(2) If two electrons occupy two different orbitals, their spins will be parallel in the ground state. Hund's rules states that the electrons attain positions as far apart as possible which minimizes the repulsion obtained from interelectronic forces.

To solve this problem one must :

(1) Find the total number of electrons within the atom.

(2) Determine the number of valence electrons.

(3) Find the number of electrons in the highest equivalent energy orbital.

The total number of electrons in an atom is equal to that atom's atomic number.

Thus, Si has 14 electrons
P has 15 electrons
S has has 16 electrons
Cl has 17 electrons
Ar has 18 electrons.

Next, from the orbital configuration :

$_{14}Si$	$1s^2 2s^2 2p^6 3s^2$	$3p^2$
$_{15}P$	$1s^2 2s^2 2p^6 3s^2$	$3p^3$
$_{16}S$	$1s^2 2s^2 2p^6 3s^2$	$3p^4$
$_{17}Cl$	$1s^2 2s^2 2p^6 3s^2$	$3p^5$
$_{18}Ar$	$1s^2 2s^2 2p^6 3s^2$	$3p^6$

One knows that the highest equivalent energy orbital is the 3p orbital. The number of electrons in this orbital increases by 1 starting with 2 for Si, then, 3 for P, 4 for S, 5 for Cl, and 6 Ar. Thus, Ar closes this orbital. Using Hund's rules, the electron configurations can be written as shown in the accompanying figure.

Q. 11 : Under ordinary conditions of temperature and pressure, nitrogen gas exists as diatomic molecules (N_2). As a result of its electronic configuration, N_2 is very inert and requires extreme conditions before it will react with any other species. What is the electronic ground state configuration of a nitrogen atom (atomic number = 7) ?

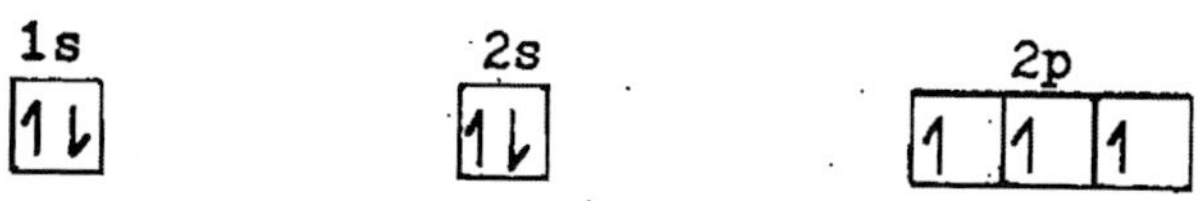

Solution : The solution to this problem involves the application of the Pauli Exclusion Principle and of Hund's Rule.

Since the atomic number of nitrogen is 7 and it is a neutral atom, there are 7 electrons to place in atomic orbitals. The two orbitals of lowest energy are the 1s and the 2s. These are filled according to the Pauli Exclusion Principle–2 electrons of unpaired spin in each orbital. The orbitals with the lowest energy of those remaining are the three 2p orbitals, $2p_x$, $2p_y$, $2p_z$. Hund's rule dictates that the remaining three electrons enter these orbitals singly and with parallel spin.

The ground state electronic distribution for nitrogen can therefore be pictured as shown in the accompanying Figure.

Q. 12 : Predict the total spin for each of the following electronic configurations : (a) $1s^2 2s^1$; (b) $1s^2 2s^2 2p^3$; and (c) $1s^2 2s^2 2p^6 3s^2 3p^6 3d^5 4s^2$.

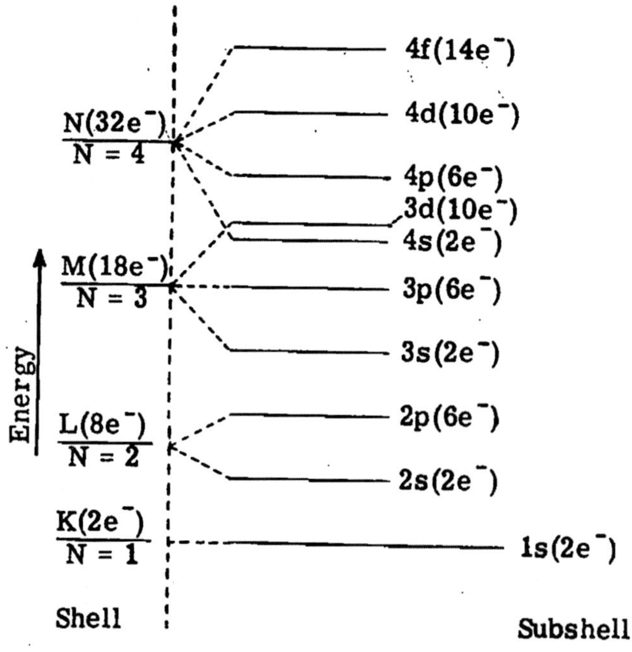

electrons	↑↓	↑↓
orbitals	1s	2s

Fig. B

electrons	↑↓	↑↓	↑	↑	↑
orbital	1s	2s	2p		

Fig. C

Solution : One determines the total spin of an atom by calculating the sum of all the s quantum numbers of the electrons. There are four quantum numbers needed to describe each electron. They are the principal quantum number n, the azimuthal quantum number ℓ, the magnetic quantum number m and the spin quantum number s.

In each orbital of the atom two electrons can be accommodated. These electrons will have opposing spins of $+\frac{1}{2}$ and $-\frac{1}{2}$. The total spin is calculated by determining the total number of unpaired electrons and multiplying this number by $\frac{1}{2}$. It is assumed that all unpaired electrons will have the same spin. Solving for the total spins : Fig. A shows the total number of electrons that can occupy each orbital.

(a) $1s^2 2s^2$. Filled s orbitals each contain 2 electrons. One writes this electronic configuration diagrammatically as shown in fig. B.

Because there are no unpaired electrons the total spin for this atom is 0.

(b) $1s^2 2s^2 2p^3$. Here the two s orbitals are filled but the p orbital is only half filled. There are three subshells in the p orbital and there will be one electron occupying each of these. The configuration of lowest energy is assumed when each of these 3 electrons has the same spin. This configuration is written diagrammatically as shown in fig. C.

This configuration contains 3 unpaired electrons. The total spin is, thus $3 \times \frac{1}{2}$ or 3/2.

(c) $1s^2 2s^2 2p^6 3s^2 3p^6 3d^5 4s^2$. From fig. A one sees that the only unfilled orbital in this configuration is the 3d. Therefore, unpaired electrons can only occur in this shell. From fig. A, one sees that the 3d orbital can contain 10 electrons. Because each subshell contains 2 electrons, there are 5 subshells in this orbital. In this configuration there are 5 electrons occupying the 3d orbital, therefore, each subshell contains 1 electron and there are 5 unpaired electrons in the configuration. The total spin of this configuration is, thus, $5 \times \frac{1}{2}$, or 5/2.

Q. 13 : Assuming all other orbitals to have zero net spin, what would be the net spin and the multiplicity of an atom having each of the following outer shell configurations : $2p^1$, $2p^3$, $2p^5$, $3d^1$, $3d^3$, $3d^5$? Rank these net spins in order of increasing paramagnetism.

Configuration	Orbital Filling	$n_\uparrow$	$n_\downarrow$	$S = \left\lvert \frac{n_\uparrow}{2} - \frac{n_\downarrow}{2} \right\rvert$	$m = 2S + 1$
p^1	p_x [↑] p_y [ ] p_z [ ]	1	0	1/2	2 (doublet)
p^3	p_x [↑] p_y [↑] p_z [↑]	3	0	3/2	4 (quartet)
p^5	p_x [↑↓] p_y [↑↓] p_z [↑]	1	0	1/2	2 (doublet)
d^1	d_{xy} [↑] d_{yz} [ ] d_{zx} [ ] $d_{x^2-y^2}$ [ ] d_{z^2} [ ]	1	0	1/2	2 (doublet)
d^3	d_{xy} [↑] d_{yz} [↑] d_{zx} [↑] $d_{x^2-y^2}$ [ ] d_{z^2} [ ]	3	0	3/2	4 (quartet)
d^5	d_{xy} [↑] d_{yz} [↑] d_{zx} [↑] $d_{x^2-y^2}$ [↑] d_{z^2} [↑]	5	0	5/2	6 (sextet)

Solution : We proceed by first defining "total spin" and "multiplicity". Let the number of electrons having "spin up" be denoted by $n_\uparrow$ and the total number of electrons having "spin down" be denoted by $n_\downarrow$. Assigning a value of +½ to spin up and –½ to spin down, the total spin s is given by the absolute value of $n_\uparrow(+½) + n_\downarrow(-½) = n_\uparrow/2 + n_\downarrow/2$. The multiplicity is then defined as m = 2s + 1. The configuration with the highest spin (highest multiplicity) will have the greatest paramagnetism. (Substances that are weakly attracted to magnets are paramagnetic.) The configuration with the lowest spin (lowest multiplicity), will be the least paramagnetic.

The p orbital can contain six electrons and the d orbital can contain 10 electrons. When these orbitals are filled, electrons are sequentially placed in each sub-orbital (e.g., the p_x or the d_{xy}) until there is one electron in each suborbital, all electrons having the same spin. Only when there is one electron in each suborbital does another electron enter a suborbital already containing an electron, this time with a spin opposite to that of the electron already in the suborbital (the spins are then "paired") in accordance with the Pauli Exclusion Principle.

In the accompanying table we show the orbital filling corresponding to each configuration. Counting $n_\uparrow$ and $n_\downarrow$, we then calculate s and m.

The configurations p^1, p^5, and d^1 have the same electron paramagnetism (all have spin ½). The next most paramagnetic configurations are p^3 and d^3 (spin 3/2) and the most paramagnetic configuration is d^5 (spin 5/2).

Q. 14 : Given the following electron configurations : (a) (Z = 11) $1s^22s^22p^6$, (b) (Z = 25) $1s^22s^22p^63s^23p^63d^5$, and (c) (Z = 29) $1s^22s^22p^63s^23p^63d^{10}$; provide the proper ionic symbol.

Solution : The ionic symbol of an atom is equal to the charge on the atom. This charge is determined by comparing the atomic number Z and the number of electrons as shown in the electronic configuration. The atomic number corresponds to the net positive charge on the nucleus and the number of electrons indicates magnitude of the negative charge of the electron cloud.

One writes the electronic configuration of hydrogen as $1s^1$, 1s indicates the atomic orbital and the superscript 1 indicates that there is one electron in the orbital. Thus, one can determine the number of electrons present by taking the sum of the superscripts. The net charge on an atom is found by adding the net negative charge (the sum of the electrons) and the net positive charge (the atomic number, which is equal to the number of protons). For hydrogen, (Z = 1) $1s^1$, the net negative charge is –1 and the net positive charge is +1, thus, the atom is neutral and no ionic symbol is used. One uses this method to determine the ionic symbols for the atoms described in the problem.

(a) (Z = 11) $1s^22s^22p^6$. The net negative charge is equal to (–2) + (–2) + (–6) or –10. The net positive charge is 11. The net charge on the atom is 11 – 10 = + 1. From the periodic table one sees that this atom is Na^+, because the atomic number of sodium is 11.

(b) (Z = 25) $1s^22s^22p^63s^23p^63d^5$. Total number of electrons in configuration is 2 + 2 + 6 + 2 + 6 + 5 = 23. The net charge on the atom is, then 25 – 23 or +2. From the periodic table one can determine that this atom is Mn^{+2}.

(c) (Z = 29) $1s^22s^22p^63s^23p^63d^{10}$. total number of electrons in configuration = 2 + 2 + 6 + 2 + 6 + 10 = 28. Z = 29 = number of electrons in a neutral atom. Difference : 29 – 28 = +1. The atom is Cu^{+1}.

Q. 15 : Assume that 90% of the electron density is representative of the volume of the atom. the following atomic radii have been obtained for 7 elements across the second period of the periodic table and for six elements down through the first family :

H						
0.37						
Li	Be	B	C	N	O	F
1.23	.89	.80	.77	.74	.74	.72
Na						
1.57						
K						
2.03						
Rb						
2.16						
Cs						
2.35						

Explain the observed trends of these atomic radii.

Solution : Atomic radii increase going down a family, but decrease going across a period. To explain this, one must consider the change in the elements in going across a period or down a family and see how it could affect atomic radii. In moving across a period, the atomic number increases, which means the nuclear positive charge increases. But the principal quantum number of the outiside electrons remains constant. This means that the outer electrons, assuming all other factors, are constant, will remain at the same distance from the nucleus.

Not all factors are constant, though. With increasing atomic number, the nuclear positive charge increases, and, thus, the force pulling the electrons towards the nucleus increases. Therefore, the atomic radii decreases as one moves across a period. Moving down a family presents a similar case of increasing atomic number. But, here, the principal quantum number is also increasing, which means the outermost electrons are farther and farther away from the nucleus. The principal quantum number has a stronger effect than atomic number in determining atomic radii. Thus, the radii increase going down a family.

3

Atomic Spectroscopy

Q. 1 : Calculate the energy (in ergs) of a single quantum of each of the following : (a) 0.5 Å X-ray, (b) a radio wave of 900 kilocycles/ sec. 10^{-8} cm = 1 Å, h = 6.626×10^{-27} erg – sec, c = 2.99×10^{10} cm/ sec and 900 kilocycles/sec = 9 × 105 sec^{-1}.

Solution : The energy (E) of a single quantum can be claculated using the formula E = hν, where h = Planck's constant and ν = frequency. Sometimes the frequency is not known, but the wavelength, λ, is. λ and ν are related via the expression $\lambda\nu = c$, where c = speed of light. Thus, if one solves for ν and then for E.

Proceed as follows :

(a) It is given that the wavelength of the X-ray = .5 Å or

$$.5\text{Å}\left(\frac{10^{-8}}{\text{Å}}\text{cm}\right) = .5\times10^{-8}\text{ cm, thus } \lambda = .5\times10^{-8}\text{ cm.}$$

and $$\nu = \frac{c}{\lambda} = \frac{2.99\times10^{10}\text{cm/s}}{.5\times10^{-8}\text{cm}} = 5.98\times10^{18}\text{ sec}^{-1}.$$

Solving for E :

$$E = h\nu = (6.626\times10^{-27}\text{ erg–sec})(5.98\times10^{18}\text{ sec}^{-1})$$
$$= 4.0\times10^{-8}\text{ erg.}$$

(b) Use E = hν. One is given the frequency as 900 kilocycles/ sec = 9×10^5 sec^{-1}. Thus,

$$E = (6.626\times10^{-27}\text{ erg–sec})(9\times10^{5}\text{ sec}^{-1})$$
$$= 5.96\times10^{-21}\text{ erg.}$$

Q. 2 : How much energy is emitted by Avogadro number of atoms, if they each emit a light wave of 400 nm wavelength ?

Solution : Energy is related to wavelength in the following equation, $E = hc/\lambda$, where E is energy, h Planck's constant (6.626×10^{-34} Js), c the speed of light (3.0×10^{8} m/sec) and λ the wavelength. Using this equation, one calculates the energy emitted by each atom. To find the total amount of energy produced by Avogadro's number of atoms, multiply the amount of energy one atom emits by 6.02×10^{23}.

Solving for E : 1 nm = 10^{-9} m.

$$E = \frac{hc}{\lambda} = \frac{(6.626\times10^{-34}\,\text{Js})\ (3.0\times10^{8}\,\text{m/s})}{(400\times10^{-9}\,\text{m})}$$

$$= 4.97\times10^{-19}\ \text{J/atom}$$

Thus, the total amount of energy emitted is

$E = (4.97 \times 10^{-19}$ J/atom) $(6.02 \times 10^{23}$ atoms/mole)

$= 2.98 \times 10^{5}$ J.

Q. 3 : The sodium D-lines, two very distinct lines of yellow light emitted by heated sodium metal, are often used in the calibration of spectrometers. The wavelength of one of these lines is 5890 Å. How much energy does an electron emit (or absorb) in undergoing the electronic transition associated with this line ?

Solution : To solve this problem we will combine two relationships : $\nu = c/\lambda$, where ν is the frequency, λ the wavelength, and c the speed of light; and Planck's relationship, $E = h\nu$, where E is the energy and h is Planck's constant, 6.62×10^{-27} erg–sec. Then, $E = h\nu = hc/\lambda$. Since 1 Å = 10^{-8} cm, 5890 Å = 5890 Å × 10^{-8} cm/Å = 5.890×10^{-5} cm. The energy associated with the 5890 Å sodium D-line is then

$$E = \frac{hc}{\lambda} = \frac{6.62\times10^{-27}\,\text{erg}-\text{sec}\times3.0\times10^{10}\,\text{cm/sec}}{5.89\times10^{-5}\,\text{cm}}$$

$$= 3.37\times10^{12}\ \text{erg.}$$

Q. 4 : In order for a photon of light incident upon metallic potassium to eliminate and electron from it, the photon must have a minimum energy of 4.33 eV (photoelectric work function for potassium). What is the wavelength of a photon of this energy ? Assume h = 6.626×10^{-27} erg–sec and c = 2.998×10^{10} cm/sec.

Solution : This problem illustrates two fundamental

relationships. The first is that the product of wavelength λ and frequency ν is the speed of light, $c = h\nu$ (or $\nu = c/\lambda$). The second is the relationship between frequency and energy E, $E = h\nu$, where h is Planck's constant. Combining these equations we obtain

$$E = h\nu = h\,(c/\lambda) = hc/\lambda, \text{ or } \lambda = hc/E.$$

Before applying this equation, we must convert the energy E from electron volts (eV) to the more convenient unit of energy, the erg. Since 1 eV = 1.602×10^{-12} erg,

$$E = 4.33 \text{ eV} = 4.33 \text{ eV} \times 1.602 \times 10^{-12} \text{ erg/eV}$$
$$= 6.94 \times 10^{-12} \text{ erg.}$$

The wavelength of light corresponding to 4.33 eV is then

$$\lambda = \frac{hc}{E} = \frac{6.626\times10^{-27}\,\text{erg}-\text{sec}\times2.998\times10^{10}\,\text{cm/sec}}{6.94\times10^{-12}\,\text{erg}}$$

$$= 2.864\times10^{-5}\,\text{cm} = 2.864\times10^{-5}\,\text{cm}\times10^{8}\,\text{Å/cm}$$

$$= 2864\text{Å}.$$

Q. 5 : Calculate the number of electrons that can be removed from metallic cesium with the energy required to remove one electron from an isolated cesium atom. Assume the following: the ionization potential of Cs = 3.89 eV, h = 6.63×10^{-34} J sec, and Cs metal is sensitive to red light of 700 nm.

Solution : This problem entails the understanding of the photoelectric effect. The photoelectric effect states that when one shines light of a given frequency on a metallic material, electrons will be emitted. The equation : $E = hv$ (where E = energy, v = frequency, and h = Planck's constant) determines the amount of energy produced by a given frequency of light. The ionization potential is the amount of energy needed to cause the electrons to be ejected from the metal. To solve this problem, therefore, one must calculate how much energy is present in light with a wavelength of 700 nm and then compare this to the ionization potential of a Cs atom.

It is given that the wavelength of the light is 700 nm. The frequency of this light is equal to speed of light/wavelength = $(3.00 \times 10^{8}$ m/s$)/700 \times 10^{-9}$ m. Thus,

$$E = (6.63\times10^{-34}\,\text{J sec})\frac{(3.00\times10^{8}\,\text{m/s})}{700\times10^{-9}\,\text{m}}$$

$$= 2.84\times10^{-19}\,\text{J/photon}$$

1 eV = 1.60 × 10^{-19} J. The ionization energy is 3.89 eV or (3.89 eV) (1.60 × 10^{-19} J/ev) = 6.22 × 10^{-19} J.

Because Cs is sensitive to light with a wavelength of 700 nm, one photon of this light flashed on the metal Cs will cause 1 electron to be emitted from the Cs. Therefore, the number of electrons emitted when 6.22 × 10^{-19} J, (the ionization potential of an isolated Cs atom),is added to the metallic system is found by dividing 6.22 × 10^{-19} J by 2.84 × 10^{-19} J

$$\text{No. of electrons emitted} = \frac{6.22\times10^{-19}\,\text{J}}{2.84\times10^{-19}\,\text{J}} = 2.2 \text{ electrons}$$

Therefore, the maximum number of electrons that can be emitted is 2.

Q. 6 : Whate wavelength of light is needed to excite an electron in a 0.2 nanometer (1 nm = 10^{-9} m) box from the ground state to the second excited state ? What wavelength of light is emitted when the same electron falls from the second excited state to the first exicted state ?

Solution : The solution to this problem requires the application of two equations, one to determine the energies of the various states, from which may be obtained the energy of a transition between states, and one to relate energy to wavelength. The first equation is the energy of a particle-in-a-box,

$$E_n = \frac{n^2h^2}{8mL^2},$$

where n is the principal quantum number (n = 1, 2, 3, ...), h is Planck's constant (6.626 × 10^{-27} erg–sec = 6.626 × 10^{-34} J–sec), m is the mass of the particle, and L is the length of the box. For an electron in a 0.2 nm

$$= 0.2 \times 10^{-9}\text{ m} = 2 \times 10^{-10}\text{ m box,}$$

$$E_n = \frac{n^2h^2}{8mL^2} = \frac{n^2(6.626\times10^{-34}\,\text{J}-\text{sec})^2}{8\times9.11\times10^{-31}\,\text{kg}\times(2\times10^{-10}\,\text{m})^2}$$

$$= 1.506\times10^{-18}\,\text{J}\times n^2.$$

The second equation relates the energy difference, ΔE, between two states and the wavelength λ : ΔE = hc\λ, or λ = hc\ΔE, where c is the speed of light.

In the following table, we calculate the energies corresponding to the ground state (n = 1), the first excited state (n = 2), and the second excited state (n = 3).

n	$E_n = n^2 \times 1.506 \times 10^{-18}$ J
1	$E_1 = (1)^2 \times 1.506 \times 10^{-18}$ J = 1.506×10^{-18} J
2	$E_2 = (2)^2 \times 1.506 \times 10^{-18}$ J = 6.024×10^{-18} J
3	$E_3 = (3)^2 \times 1.506 \times 10^{-18}$ J = 1.355×10^{-17} J

In going from the ground state to the second excited state, the electron must absorb a quantum of light of energy $\Delta E = E_3 - E_1 = 1.355 \times 10^{-17}\text{ J} - 1.506 \times 10^{-18}\text{ J} = 1.204 \times 10^{-17}$ J. The corresponding wavelength is

$$\lambda = \frac{hc}{\Delta E} = \frac{6.626\times10^{-34}\,J\times3\times10^{8}\ m/sec}{1.204\times10^{-17}\,J}$$
$$= 16.510\times10^{-9}\,m = 16.510nm.$$

In falling from the second excited state to the first excited state the electron will emit a quantum of light of eneryg $\Delta E = E_3 - E_2 = 1.355 \times 10^{-17}\text{ J} - 6.024 \times 10^{-18}\text{ J} = 7.526 \times 10^{-18}$ J. This corresponds to a wavelength of

$$\lambda = \frac{hc}{\Delta E} = \frac{6.626\times10^{-34}\,J\times3\times10^{8}\,m/sec}{7.526\times10^{-18}\,J}$$
$$= 26.412\times10^{-9}\,m = 26.412\,nm.$$

Q. 7 : Consider an atom containing one electron and having an atomic number Z. It is observed that light of a wavelength of 0.1 nm will excite this atom from the ground state to infinite ionization. Determine the value of Z.

Solution : We know the wavelenght λ of the exciting light and the quantum numbers of the ground state (n_g = 1) and the ionized state ($n_i = \infty$). We must determine Z, the number of protons in the nucleus of the atom.

There is a relationship to find Z using n_g and n_i. It is provided by the Bohr theory of the atom. It is :

$$\frac{c}{\lambda} = \frac{Z^2e^2}{2ha_0}\left(\frac{1}{n_g^2} - \frac{1}{n_i^2}\right),$$

where c is the speed of light ($c = 2.9969 \times 10^{10}$ cm/sec), e is the electronic charge ($e = 4.80 \times 10^{-10}$ esu), h is Planck's constant ($h = 6.626\times10^{-27}$ erg-sec), and a_0 is the Bohr radius ($a_0 = 0.05292 \times 10^{-7}$ cm). If we combine some of these constants, we arrive at the value $e^2/2ha_0 = 3.290 \times 10^{15}$ sec^{-1}. Hence, our expression becomes

$$\frac{c}{\lambda} = Z^2\left(\frac{e^2}{2ha_0}\right)\left(\frac{1}{n_g^2}-\frac{1}{n_i^2}\right)$$

$$= Z^2 \times 3.290\times10^{15}\ \text{sec}^{-1}\left(\frac{1}{n_g^2}-\frac{1}{n_i^2}\right).$$

To solve this problem, we must solve this equation for Z and substitute c, n_g, n_i, and λ ($\lambda = 0.1$ nm $= 0.1 \times 10^{-9}$ m$=10^{-10}$ m$=10^{-8}$ cm) into the resulting equation.

Solving for Z, we obtain

$$Z^2 = \frac{c}{\lambda}\left[3.290\times10^{15}\,\text{sec}^{-1}\left(\frac{1}{n_g^2}-\frac{1}{n_i^2}\right)\right]^{-1}, \quad \text{or}$$

$$Z = \left(\frac{c}{\lambda\times3.290\times10^{15}\,\text{sec}^{-1}\left(\frac{1}{n_g^2}-\frac{1}{n_i^2}\right)}\right)^{1/2}$$

Substituting the values of c, λ, n_g, and n_i into this expression yields

$$Z = \left(\frac{2.9969\times10^{10}\,\text{cm / sec}}{10^{-8}\,\text{cm}\times3.290\times10^{15}\,\text{sec}^{-1}\left(\frac{1}{1}-\frac{1}{\infty}\right)}\right)^{1/2} \cong 30,$$

where we used the equality $1/\infty = 0$. Hence, this atom must have 30 protons.

Q. 8 : The Rydberg-Ritz equation governing the spectral lines of hydrogen is $\frac{1}{\lambda} = R\left(\frac{1}{n_{1l}^2}-\frac{1}{n_2^2}\right)$, where R is the Rydberg constant,

n_1 indexes the series under consideration ($n_1 = 1$ for the Lyman series, $n_1 = 2$ for the Balmer series, $n_1 = 3$ for the Paschen series), $n_2 = n_1 + 1, n_1 + 2, n_1 + 3, ...$ indexes the successive lines in a series, and λ is the wavelength of the line corresponding to index n_2. Thus, for the Lyman series, $n_1 = 1$ and the first two lines are 1215.56 Å ($n_2 = n_1 + 1 = 2$) and 1025.83 Å ($n_2 = n_1 + 2 = 3$). Using these two lines, calculate two separate values of the Rydberg constant. The actual value of this constant is R = 109678 cm^{-1}.

Solution : The first thing to do is to convert the wavelengths from Å to more manageable units, i.e., centimeters. Using the relationship 1 Å = 10^{-8} cm, the first two Lyman lines are 1215.56 Å = 1215.56 × 10^{-8} cm for $n_2 = 2$, and 1025.83 Å = 1025.83 × 10^{-8} cm for $n_2 = 3$. Solving the Rydberg-Ritz equation for R, one obtains :

$$R = \left(\lambda\left(\frac{1}{n_1^2} - \frac{1}{n_2^2}\right)\right)^{-1},$$

For the first line,

$$R = \left(\lambda\left(\frac{1}{n_1^2} - \frac{1}{n_2^2}\right)\right)^{-1} = \left(1215.56\times10^{-8}\,\text{cm}\left(\frac{1}{1^2} - \frac{1}{2^2}\right)\right)^{-1}$$

$$= 109689\ \text{cm}^{-1}$$

and for the second line,

$$R = \left(\lambda\left(\frac{1}{n_1^2} - \frac{1}{n_2^2}\right)\right)^{-1} = \left(1025.83\times10^{-8}\,\text{cm}\left(\frac{1}{1^2} - \frac{1}{3^2}\right)\right)^{-1}$$

$$= 109667\ \text{cm}^{-1}.$$

The first of these is 0.0100% greater than the true value, and the second is 0.0100% less than the true value.

Q. 9 : Wave number ($\bar{\nu}$) are the reciprocals of wavelengths, λ, and are given by the expression $\bar{\nu} = 1/\lambda$. For the hydrogen atom, the Bohr theory predicts that the wave number for the emission line associated with an electronic transition from the energy level having principal quantum number n_2 to that with principal quantum number n_1 is

$$\bar{\nu} = R_H\left(\frac{1}{n_1^2} - \frac{1}{n_2^2}\right),$$

where R_H is the Rydberg constant. In what region of the electromagnetic spectrum would there appear a spectral line resulting from the transition from the tenth to the fifth electronic level in hydrogen ?

Solution : This problem can be solved by applying the Rydberg formula used for determining the wavenumbers of the spectral lines.

The Rydberg constant is 1.10×10^5 cm^{-1}. $n_2 = 10$ and $n_1 = 5$. Substituting:

$$\bar{\nu} = R_H\left(\frac{1}{n_1^2} - \frac{1}{n_2^2}\right) = 1.10 \times 10^5 \text{cm}^{-1}\left(\frac{1}{5^2} - \frac{1}{10^2}\right)$$

$$= 1.10 \times 10^5 \text{cm}^{-1}\left(\frac{1}{25} - \frac{1}{100}\right) = 1.10 \times 10^5 \text{cm}^{-1}(0.04 - 0.01)$$

$$= 1.10 \times 10^5 \text{cm}^{-1} \times 0.03$$

$$= 3.3 \times 10^3 \text{cm}^{-1}.$$

This line appears in the infrared region of the spectrum (approximately 20 cm^{-1} to 104 cm^{-1}).

Q. 10 : What is the maximum number emission lines one would expect to see in a molecule containing only the six electronic energy levels depicted in diagram A ?

Diagram A

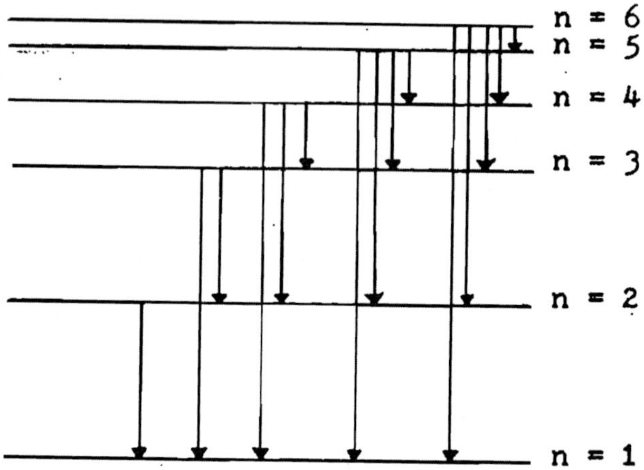

Diagram B

Solution : Emission is observed when an electron falls from an electronic energy level of higher energy to one of lower energy.

From the energy level n = 6, electrons can fall to energy levels of n = 5, 4, 3, 2, and 1. Thus, five emission lines are drawn from the energy level n = 6. By using a similar process, one determines that the maximum number of emission lines from energy levels n = 5, 4, 3, 2, and 1 are four, three, two, one, and zero, respectively. Because there are no energy levels lower than n = 1, no emission lines are obtained from it. The total number of emission lines is found by taking the sum of all the emission lines from the various energy levels. Thus, 5 + 4 + 3 + 2 + 1 + 0=15 emission lines can be expected. The possible emission processes are depicted in diagram B.

Q. 11 : Find the λ's (wavelengths) for the first four hydrogen spectral lines in the Paschen series. Find the wavelength of the series limit.

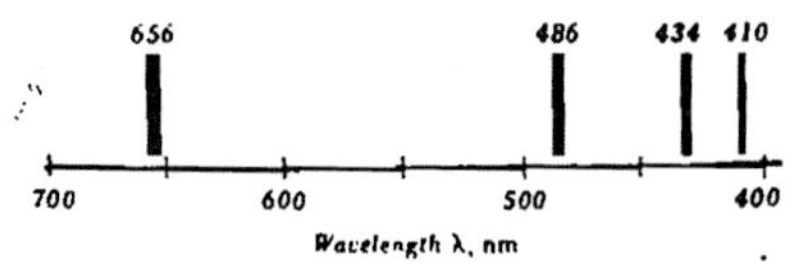

Hydrogen line spectrum. (Numbers shown are wavelengths in nanometers, or tens of angstrom units.)

Solution : The Paschen series illustrates the spectrum hydrogen exhibits in the infrared region. An example of a hydrogen line spectrum is shown in the accompanying figure. The series of lines of the Paschen series fit the equation

$$\nu = 3.29 \times 10^{15}\left(\frac{1}{9} - \frac{1}{b^2}\right),$$

where ν = frequency and b = 4, 5, 6, 7, 8, ... Thus, if you want the first four hydrogen spectral lines, you use b = 4, 5, 6, and 7. Your answer would be in terms of frequency, however, you want the wavelengths. To obtain this, employ the relation between frequency and wavelength, $c = \lambda\nu$, where c = speed of light, λ = wavelength and ν = frequency. ($c = 2.988 \times 10^{10}$ cm/sec).

Once you know ν, you can find λ. Therefore,

1st line : $b = 4,\ \nu = 3.290 \times 10^{15}\left(\frac{1}{9} - \frac{1}{16}\right)$

$= 1.599 \times 10^{14}\ \text{sec}^{-1}.$

2nd line : $b = 5,\ \nu = 3.290 \times 10^{15}\left(\frac{1}{9} - \frac{1}{25}\right)$

$= 2.340 \times 10^{14}\ \text{sec}^{-1}.$

3rd line : $b = 6,\ \nu = 3.290 \times 10^{15}\left(\frac{1}{9} - \frac{1}{36}\right)$

$= 2.742 \times 10^{14}\ \text{sec}^{-1}.$

4th line : $b = 7,\ \nu = 3.290 \times 10^{15}\left(\frac{1}{9} - \frac{1}{49}\right)$

$= 2.984 \times 10^{14}\ \text{sec}^{-1}.$

The series limit is b = infinity. If $B \to \infty$, then $\frac{1}{B^2} \to 0$. Series limit: $\nu = 3.290 \times 10^{15}\left(\frac{1}{9} - 0\right) = 3.656 \times 10^{14}$

$$\text{The wavelengths } (\lambda) = \frac{\text{speed of light (c)}}{\text{frequency } (\nu)}$$

1st line: $\lambda = 1.87 \times 10^{-4}$ cm 2n line: $\lambda = 1.28 \times 10^{-4}$ cm

3rd line: $\lambda = 1.09 \times 10^{-4}$ cm   4th line: $\lambda = 1.004 \times 10^{-4}$ cm

Series limit : $.820 \times 10^{-4}$ cm.

Q. 12 : The first two lines in the Lyman series for hydrogen are 1215.56 Å and 1025.83 Å. These lines lie in the ultraviolet region of the spectrum. For each of these lines calculate the following : (a) the corresonding energy in ergs; (b) the corresponding energy in Kcal/mole; (c) the frequency in sec^{-1}.

Solution : The problem involves the application of two basic formulas. The first is the relationship between wavelength (λ) and frequency (ν). The product of the two is the speed of light c, $\lambda\nu = c$ (or $\nu = c/\lambda$). The second is the relationship between frequency and energy (E), $E = h\nu$, where h is Planck's constant.

First calculate the frequencies for the two lines. Distinguish λ, ν, and E for these lines with the subscripts "1" and "2". thus,

$$\lambda_1 = 1215.56 \text{ Å} \quad \text{and} \quad \lambda_2 = 1025.83 \text{ Å}.$$

To express these in the more convenient units of centimeters, use the relationship 1 Å = 10^{-8} cm. Then,

$$\lambda_1 = 1215.56 \times 10^{-8} \text{ cm} \quad \text{and} \quad \lambda_2 = 1025.83 \times 10^{-8} \text{ cm}.$$

Using the first of the two formulas, one obtains :

$$\nu_1 = \frac{c}{\lambda_1} = \frac{2.9979 \times 10^{10} \text{cm/sec}}{1215.56 \times 10^{-8} \text{cm}} = 2.4663 \times 10^{15} \text{ sec}^{-1}$$

$$\nu_2 = \frac{c}{\lambda_2} = \frac{2.9979 \times 10^{10} \text{cm/sec}}{1025.83 \times 10^{-8} \text{cm}} = 2.9224 \times 10^{15} \text{ sec}^{-1}.$$

The energies for these two lines are calculated using the second of the two relationships. Planck's constant is $h = 6.626 \times 10^{-27}$ erg–sec. The units of frequency are sec^{-1}, the units of energy (found by using $E = h\nu$), will then be erg–sec × sec^{-1} = erg. To convert from ergs to Kcal/mole, first convert to joules (1 joule = 10^7 ergs), then from joules to Kcal (1 Kcal = 4.184×10^3 joules), and finally multiply by Avogadro's number (6.022×10^{23}/mole) in order to put this on a per mole basis. Hence, the conversion factor from ergs to Kcal/mole is

$$1 \text{ erg} = 1 \text{ erg} \times 1 \text{ joule}/10^7 \text{ ergs} \times 1 \text{ Kcal}/4.184 \times 10^3 \text{ joules}$$

$\times$ 6.022 $\times$ 10^{23}/mole

= 1.439 $\times 10^{13}$ Kcal/mole.

Applying the second equation one obtains:

$E_1 = h\nu_1$ = 6.626 $\times$ 10^{-27} erg–sec $\times$ 2.4663 $\times$ 10^{15} sec^{-1}

= 1.6342 $\times$ 10^{-11} erg

= 1.6342 $\times$ 10^{-11} erg $\times$ 1.439 $\times$ 10^{13} Kcal/mole–erg

=235.2 Kcal/mole.

$E_2 = h\nu_2$ = 6.626 $\times$ 10^{-27} erg–sec $\times$ 2.9224 $\times$ 10^{15} sec^{-1}

= 1.9364 $\times$ 10^{-11} erg

= 1.9364 $\times$ 10^{-11} erg $\times$ 1.439 $\times$ 10^{13} Kcal/mole–erg

=278.6 Kcal/mole.

Q. 13 : Show that in the Balmer series that the frequencies of successive lines tend toward a limiting value of cR/4, where R is the Rydberg constant and c the speed of light.

Solution : Here, one must try to relate frequency ν with c and R. The Rydberg constant, R, comes from the Rydberg Ritz equation,

$$\frac{1}{\lambda} = R\left(\frac{1}{(n_1)^2} - \frac{1}{(n_2)^2}\right),$$

where λ is wavelength. The electron falls from an energy level of n_2 to n_1. The speed of light is related to the frequency in the equation, $\nu = c/\lambda$.

These equations are combined.

$$(c)\left(\frac{1}{\lambda}\right) = cR\left(\frac{1}{(n_1)^2} - \frac{1}{(n_2)^2}\right) = \frac{c}{\lambda} = \nu$$

Thus, $$\nu = cR\left(\frac{1}{(n_1)^2} - \frac{1}{(n_2)^2}\right)$$

In the Balmer series n_1 is defined as being equal to 2. For successive lines, $n_2 \to \infty$ and therefore $1/(n_2)^2$ approaches 0. The above equation becomes

$$\nu = cR\left(\frac{1}{(2)^2} - \frac{1}{(\infty)^2}\right) = \frac{cR}{4}.$$

Q. 14 : Find the difference in characteristic virbational frequency of a diatomic molecule of mass 20 amu if (a) the mass is equally distributed between the two ends, and (b) one atom of 1 amu is at one end and an atom of 19 amu is at the other.

Solution : The vibrational frequency is frequency of the oscillation between the two atoms of the molecule. It is expressed quantitatively by

$$\nu = ½\pi\sqrt{K / M_{eff}}$$

where ν is the vibrational frequency, K a force constant, and M_{eff} the effective mass of the molecule.

$$M_{eff} = \frac{M_A M_B}{M_A + M_B}$$

where M_A is the mass of one atom and M_B is the other.

(a) If the mass is equally distributed between both atoms, each weighs 10 amu.

$$M_{eff} = \frac{10\times10}{10+10} = 5$$

Assume that K =1.

$$v = \frac{1}{2\pi}\sqrt{\frac{1}{5}} = 0.0711$$

(b) $$M_{eff} = \frac{1\times19}{20} = \frac{19}{20}$$

$$\nu = \frac{1}{2\pi}\sqrt{\frac{19}{20}} = 0.1632$$

Therefore, the atoms in the molecule described in b are vibrating more quickly than those in molecule a.

Q. 15 : Calculate the characteristic virbational frequency of the lowest vibration state for H_2. Assume k = 5.1 newtons/cm and h = 6.63 × $1^{-3}4$ J–sec.

Solution : Picture a diatomic molecule, such as H_2, as two balls connected by a spring. The spring represents the chemical bond. When the two balls (atoms) are pulled apart the string acts as a restoring force to bring them back together to the equilibrium separation. When they are compressed, the string tends to force them apart. Such a system, in which the restoring force is proportional to the amount of distrortion, is called a harmonic oscillaotr. The vibrational frequency of the spring going back and forth is given by the formula

$$\nu = \frac{1}{2\pi}\sqrt{k / M_{eff}}$$

where ν = characteristic vibrational frequency of the particular bond, k = force constant and M_{eff} = effective mass for the vibrating atoms:

$$M_{eff} = \frac{M_A M_B}{M_A + M_B}$$

If the masses of the two atoms are equal, M_{eff} becomes

$$\frac{M^2}{2M} = ½\ M.$$

To find the characteristic vibration frequency, use this formula.

To find the M_{eff}, find the mass of one hydrogen atom. The number of grams in one mole (molecular weight) is one. There are 6.02×10^{23} atoms of H in one mole. Thus, the mass of one atom is transition is from $n_2 = 2$ to $n_1 = 1$. One solved for λ by using the Rydberg-Ritz equation.

$$\frac{1}{\lambda} = (109{,}678\ \text{cm}^{-1})\left(\frac{1}{(1)^2} - \frac{1}{(2)^2}\right)$$

$$\frac{1}{\lambda} = 82258\ \text{cm}^{-1}$$

$$\lambda = \frac{1}{82258\ \text{cm}^{-1}} = 1.21 \times 10^{-5}\ \text{cm}.$$

Wavelengths are usually expressed in angstroms. $1\ \text{Å} = 10^{-8}$ cm.

$$(1.21 \times 10^{-5}\text{cm})\frac{1\text{Å}}{10^{-8}\text{cm}} = 1210\text{Å}$$

Paschen series : The first transition in the Paschen series is from $n_2 = 4$ to $n_1 = 3$ because the base energy level in this series is 3. Here again, one uses the Rydberg-Ritz equation to solve for λ.

$$\frac{1}{\lambda} = 109678\ \text{cm}^{-1}\left(\frac{1}{(3)^2} - \frac{1}{(4)^2}\right) = 5332\ \text{cm}^{-1}$$

$$\lambda = \frac{1}{5332\ \text{cm}^{-1}} = 1.87\times10^{-4}\ \text{cm}.$$

Converting to angstroms :

$$1.87\times10^{-4}\ \text{cm}\times\frac{1\text{Å}}{10^{-8}\ \text{cm}} = 18700\text{Å}.$$

Q. 16 : A gaseous substance exhibits a line spectrum, whereas a liquid or a solid substance exhibits a continuous spectrum. Give a qualitative explanation for these phenomena.

Solution : The basic difference between gases and the condensed phases of matter, which gives rise to these different phenomena, is that in gases the atoms or molecules are sufficiently far apart to have a negligible interaction, whereas in liquids and solids, the interaction between the constituent atoms and molecules is appreciable.

In the gas phase, the molecules are so far apart that they may be considered to be isolated from one another and therefore behave independently of one another. Thus, the spectrum arises only from electronic, vibrational, or rotational transitions of individual atoms or molecules, and from any collective behavior. But since these transitions are well defined energetically, the spectrum exhibits sharp lines.

In a liquid or solid, the atoms or molecules are sufficiently close together to affect each other's energy levels. That is, in addition to the individual energy levels of each atom or molecule, there exist energy levels for the collective excitation of groups of atoms or molecules. In general, the interaction of one atom or molecule with another decreases with internuclear distance or intermolecular distance in a continuous fashion, so that a continuous spectrum is observed.

4

Quantum Chemistry

Q. 1 : Which of the following operators is Hermitian; d/dx, i d/dx, d^2/dx^2?

Solution : An operator A is said to be hermitian (or self-adjoint) if it has the following property :

$$\int \alpha^* A\beta dx = \int \beta(A\alpha)^* dx$$

with respect to all the functions.

(a) for the operator $\frac{d}{dx}$,

$$\left(\frac{d}{dx}\right) = \int_{-\infty}^{\infty} \alpha^* \frac{d\alpha}{dx} dx$$

$$= \alpha^* \alpha \Big|_{-\infty}^{\infty} - \int_{-\infty}^{\infty} \alpha \frac{d\alpha^*}{dx} dx$$

$$= -\int_{-\infty}^{\infty} \alpha \frac{d\alpha^*}{dx} dx$$

Thus the operator (d/dx) is non-Hermitian.

(b) for $i\frac{d}{dx}$,

$$\left(i\frac{d}{dx}\right) = \int_{-\infty}^{\infty} i\alpha^* \frac{d\alpha}{dx} dx$$

$$= i\alpha^*\alpha\Big|_{-\infty}^{\infty} - \int_{-\infty}^{\infty} i\alpha \frac{d\alpha^*}{dx} dx$$

$$= -\int_{-\infty}^{\infty} i\alpha \frac{d\alpha^*}{dx} dx$$

(c) for $\frac{d^2}{dx^2}$,

$$\left(\frac{d}{dx^2}\right) = \int_{-\infty}^{\infty} \alpha^* \frac{d^2\alpha}{dx^2} dx$$

$$= \alpha^* \frac{d\alpha}{dx}\Big|_{-\infty}^{\infty} - \int_{-\infty}^{\infty} \frac{d\alpha}{dx} \frac{d\alpha^*}{dx} dx$$

$$= -\int_{-\infty}^{\infty} \frac{d\alpha}{dx} \frac{d\alpha^*}{dx} dx$$

$$= -\frac{d\alpha^*}{dx}\Big|_{-\infty}^{\infty} + \int_{-\infty}^{\infty} \alpha \frac{d^2\alpha^*}{dx^2} dx$$

$$= \int_{-\infty}^{\infty} \alpha \frac{d^2\alpha^*}{dx^2} = \left(\frac{d^2}{dx^2}\right)$$

The above results show that only b and c are Hermitian.

Q. 2 : A particle of mass m is confined to a one-dimensional box with the origin at the center of the box. The box extends from –L to +L. The potential energy is

$$V(x) = \begin{cases} 0, & -L \le x \le L \\ \infty, & |x| > L \end{cases}$$

(a) Write the Schrödinger equation for this problem showing separate equations for the inside and the outside of the box.

(b) Assume a solution (inside the box) of the form

$$\psi(x) = A \sin(cx) + B \cos(cx).$$

Give a rule that determines all the possible values of c, and for each possible c give the conditions on A and B that make ψ a satisfactory solution.

(c) Express the energy in terms of c.

Solution : One of the basic functions of quantum mechanics is that the motion of a partcile, that is, its position as a function of space and time, is exactly described by means of a probability density function $\rho(x, y, z, t)$. By definition, the probability of finding the particle at a time t_0 in a volume element dv, surrounding the point (x_0, y_0, z_0), is given by

$$P = \rho\,(x_0, y_0, z_0, t_0)\,dv.$$

But in quantum mechanics the probability density function ρ is derived from another function ψ, which is known as the wave function. The wave function is obatined as the solution of a differential equation that is known as the Schrödinger equation. If the motion of the particle is restricted to only one dimension the Schrödinger equation takes the form

$$-\frac{\hbar^2}{2m}\frac{d^2\psi}{dx^2} + V(x)\psi = E\psi$$

where m = mass of the particle

$\hbar = \frac{h}{2\pi}$. Here h is Planck's constant.

E = time independent total energy

ψ = wave function

V = the electrostatic interaction between two particles with charges z_1 and z_2 given by $V = z_1z_2/r$. Here r is the distance between the particles.

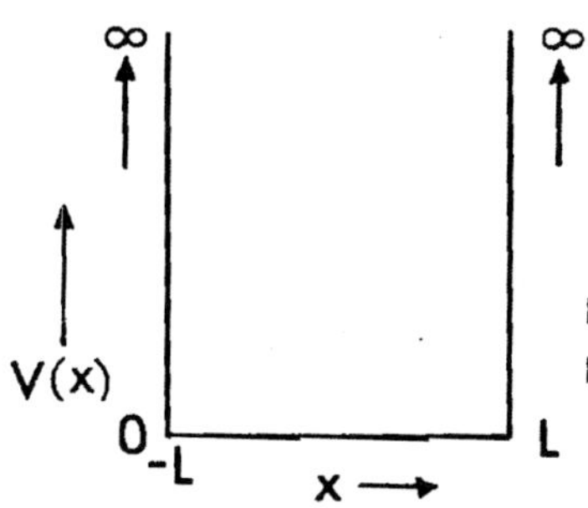

POTENTIAL FUNCTION OF THE PARTICLE IN A BOX.

Fig. 1

With reference to equation (1), a one-dimensional box is defined by a potential which is zero for $-L \le x \le L$ and which is infinite elsewhere (see Fig. 1).

Algebraically, the potential is given by

$$V(x) = \infty \qquad x < -L$$
$$V(x) = 0 \qquad 0 \le x \le L$$
$$V(x) = \infty \qquad L < x$$

The particle cannot obviously move outside the box and

$$\psi(x) = 0 \qquad x < -L \quad \text{and} \quad x > L$$

Then the Schrödinger equation becomes

$$= -\frac{\hbar^2}{2m}\frac{d^2\psi}{dx^2} + V\psi = E\psi.$$

(a) The Schrödinger equation for the region $x = -L$ and $x = L$ (inside the box) reduces to

$$\left(\frac{-\hbar^2}{2m}\right)\frac{d^2\psi}{dx^2} = E\psi \tag{1}$$

since the potential is defined to be zero in this region.

After substituting in the value of $\hbar$, the Schrödinger equation for the inside of the box is

$$\left(-\frac{h^2}{8\pi^2 m}\right)\frac{d^2\psi}{dx^2} = E\psi \tag{2}$$

For the outside, it is

$$\left(\frac{-h^2}{8\pi^2 m}\right)\frac{d^2\psi}{dx^2} + V\psi = E\psi$$

or

$$\left(\frac{-h^2}{8\pi^2 m}\right)\frac{d^2\psi}{dx^2} + \infty\psi = E\psi$$

where $V = \infty$ for the region $|x| > L$

Then $\psi = 0$ outside the box.

(b) For the equation dealing with the inside of the box a possible solution is of the form

$$\psi(x) = A \sin cx + B \cos cx$$

For continuity of inside and outside functions

$$\psi(-L) = -A \sin cL + B \cos cL = 0 \quad (3)$$

$$\psi(+L) = A \sin cL + B \cos cL = 0 \quad (4)$$

subtract equation (3) from equation (4)

$$2A \sin cL = 0 \quad (5)$$

Therefore, either A = 0 or sin cL = 0, which makes

$$cL = n\pi \text{ for } n = \pm 1, \pm 2, \ldots$$

from which

$$c = \frac{\pi n}{L}$$

Therefore, the wavefunction is given by

$$\psi = A \sin \frac{n\pi x}{L}.$$

Addition of equation (3) to equation (4) gives

$$2B \cos cL = 0$$

either B = 0 or cos cL = 0 for n = 0, ±1, ±2, ... which makes

$$cL = \left(n - \frac{1}{2}\right)\pi$$

and

$$c = \frac{\left(n - \frac{1}{2}\right)\pi}{L} = \frac{(2n-1)\pi}{2L}$$

Therefore,

$$\psi = B \cos\left[\frac{(2n-1)\pi x}{2L}\right]$$

(c)

$$E\psi = \left(\frac{-h^2}{8\pi^2 m}\right)\frac{d^2\psi}{dx^2}.$$

Also

$$\frac{d^2\psi}{dx^2} = -c^2\psi$$

$$\therefore \quad E\psi = \left(\frac{-h^2}{8\pi^2 m}\right)(-c^2\psi)$$

and

$$E = \frac{h^2c^2}{8\pi^2 m}.$$

Q. 3 : Determine the energy levels and wave functions of a particle in the potential well shown by sovling the Schrödinger equation.

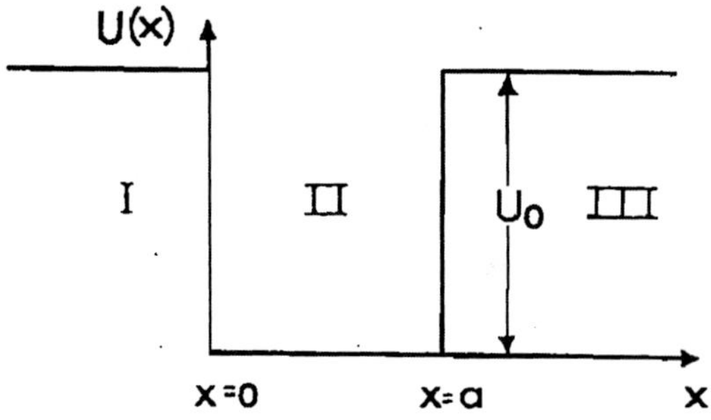

Fig. 1

Solution : In region I where $x \leq 0$, $U = U_0$. In region II, $0, \leq x \leq a$ and $U = 0$ and in region III $x > a$ and $U = U_0$. The Schrödinger equation is written as

$$\frac{d^2\psi}{dx^2} + \frac{8\pi^2 m}{h^2}(E - U)\psi = 0 \qquad (1)$$

where ψ = the amplitude function called the wave function. Equation 1 would be solved for ψ in each of the three regions of figure 1 and the constants evaluated by the requirements that ψ and $d\psi/dx$ and therefore $\left(\frac{1}{\psi}\right)(d\psi / dx)$ must be continuous at the boundaries.

Writing equation (1) for regions I and III where $U = U_0$ gives

$$\frac{d^2\psi}{dx^2} + \frac{8\pi^2 m}{h^2}(E - U_0)\psi = 0 \qquad (2)$$

In region III where $U = 0$, equation (1) becomes

$$\frac{d^2\psi}{dx^2} + \frac{8\pi^2 m}{h^2}E\psi = 0 \qquad (3)$$

Solve for ψI in region I, by making the following substitution.

Let

$$k_1^2 = \frac{8\pi^2 m}{h^2}(U_0 - E) \quad (4)$$

Substitute equation (4) into equation (2)

$$\frac{d^2\psi}{dx_2} - k_1^2 = 0$$

from which

$$\psi = Ae^{k_1x} + Be^{-k_1x} \quad (5)$$

where A and B are constants. When $x \to -\infty$ the term Be^{-k_1x} increases to infinity and B = 0 for region I. Therefore

$$\psi_I = Ae^{k_1x}$$

For region III, again

$$\psi_I = Ae^{k_1x} + Be^{-k_1x},$$

and as $x \to \infty$, Ae^{k_1x} increases to infinity and A = 0. Therefore $\psi_{III} = Be^{-k_1x}$. Define $B' = Be^{-k_1a}$, thus

$$\psi_{III} = B'e^{-k_1(x-a)} \quad (6)$$

To solve for ψ in region II, the following substitution is made. Let

$$k_2^2 = +\frac{8\pi^2 mE}{h^2} \quad (7)$$

Substitute equation (7) into equation (3) to give

$$\frac{d^2\psi}{dx_2} + k_2^2\psi = 0$$

where ψ_{II} = C sin $(k_2x + \delta)$, C and δ = constants.

To determine the energy levels of the particles some boundary conditions would be applied to the different regions of the potential well. For regions I and II the boundary conditions are

$$\psi_I = \psi_{II} \quad \text{at} \quad x = 0$$

$$\therefore \quad \frac{d\psi_I}{dx} = \frac{d\psi_{II}}{dx} \quad \text{at} \quad x = 0 \quad (8)$$

from which

$$\psi_I)_{x=0} = \psi_{II})_{x=0} = C \sin \delta \text{ and } A = C \sin \delta \quad (9)$$

Thus

$$\left.\frac{d\psi_I}{dx}\right)_{x=0} = k_1 A e^{k_1 x}$$

$$= k_1 A \text{ since } x = 0$$

Also

$$\left.\frac{d\psi_{II}}{dx}\right)_{x=0} = 0 = k_2 \cos(k_2 x + \delta)$$

$$= k_2 C \cos \delta \text{ since } x = 0$$

From equation (8)

$$k_1 A = k_2 C \cos \delta \quad (10)$$

Applying boundary conditions on regions II and III gives

$$\psi_{II} = \psi_{III} \quad \text{at} \quad x = a$$

or

$$\frac{d\psi_{II}}{dx} = \frac{d\psi_{III}}{dx} \quad \text{at} \quad x = a$$

$$\psi_{II})_{x=a} = K_2 C \cos(K_2 a + \delta)$$

$$\psi_{III})_{x=a} = C \sin(K_2 a + \delta)$$

Define $B' = C \sin (K_2 a + \delta)$ (11)

$$\therefore \quad \psi_{III})_{x=a} = B'$$

The derivative of the wave function in regions II and III will now be evaluated at point $x = a$. Thus

$$\left.\frac{d\psi_{II}}{dx}\right)_{x=a} = K_2 C \cos(K_2 a + \delta)$$

and

$$\left.\frac{d\psi_{III}}{dx}\right)_{x=a} = -K_1 B' e^{-K_1(x-a)}\Big)_{x=a}$$

$$= -K_1 B'$$

$$\therefore \quad -K_1 B' = K_2 C \cos(K_2 a + \delta) \quad (12)$$

From regions I and II boundary conditions define

$$\tan\delta = \frac{\sin\delta}{\cos\delta}$$

Using equation (9), $\sin\delta = \frac{A}{C}$ and from equation (10) $\cos\,\delta = \frac{K_1}{K_2}\,\frac{A}{C}$. Therefore,

$$\tan\delta = \frac{\frac{A}{C}}{\frac{K_1}{K_2}\frac{A}{C}}$$

$$= \frac{A}{C}\cdot\frac{K_2}{K_1}\frac{C}{A}$$

$$= \frac{K_2}{K_1}$$

Substitute for K_1 and K_2 from equations (4) and (7) gives

$$\tan\delta = \frac{\sqrt{\frac{8\pi^2 mE}{h^2}}}{\sqrt{\frac{8\pi^2(U_0 - E)}{h^2}}}$$

$$= \sqrt{\frac{E}{(U_0 - E)}} \qquad (13)$$

From regions II and III boundary conditions

$$\tan(K_2 a + \delta) = \frac{\sin(K_2 a + \delta)}{\cos(K_2 a + \delta)} \qquad (14)$$

From equation (II) $\sin\,(K_z a + \delta) = \frac{B'}{C}$ and from equation (12) $\cos(K_z a + \delta) = \frac{-K_1 B'}{K_2 C} = -\frac{K_1}{K_2}\frac{B'}{C}$. Substitute these values into equation (14) to give

$$\tan(K_2a+\delta)=\frac{\frac{B'}{C}}{-\frac{-K_1}{K_2}\frac{B'}{C}}$$

$$=-\frac{B'}{C}\frac{K_2}{K_1}\frac{C}{B'}$$

$$=-\frac{K_2}{K_1}$$

Using a trigonometry identity, define

$$\tan(K_2a+\delta)=\frac{\tan K_2a+\tan\delta}{1-\tan K_2a\tan\delta}$$

and $$\tan K_2a=\frac{2\tan\delta}{\tan^2\delta-1} \qquad (15)$$

But from equation (7)

$$K_2=\sqrt{\frac{8\pi^2mE}{h^2}}$$

and from equation (13) $\tan\delta=\sqrt{\frac{E}{U_0-E}}$ therefore equation (15) becomes

$$\tan\sqrt{\frac{8\pi^2mEa^2}{h^2}}=\frac{2\sqrt{E/(U_0-E)}}{\frac{E}{(U_0-E)}-1}$$

$$=\frac{2\sqrt{\frac{E}{U_0-E}}}{\frac{2E-U_0}{U_0-E}}$$

$$=\frac{2\sqrt{\frac{E}{U_0-E}}(U_0-E)}{2E-U_0}$$

$$=\frac{2\sqrt{E(U_0-E)}}{2E-U_0} \qquad (16)$$

Equation (16) is the energy equation with the number of bound states dependent on U_0. For example, when $U_0 >> E$, the right side of the equation approaches zero. Therefore the equation becomes

$$\tan\sqrt{\frac{8\pi^2 ma^2 E}{h^2}} = 0$$

but $\tan n\pi = 0$

$$\therefore \qquad \sqrt{\frac{8\pi^2 ma^2 E}{h^2}} = n\pi$$

Squaring both sides gives

$$\frac{8\pi^2 ma^2 E}{h^2} = n^2\pi^2$$

Solving for E

$$E = \frac{n^2 h^2}{8ma^2}.$$

Q. 4 : An electron is confined to a molecule of length 1.0 nm (10 Å, about 5 atoms long). What is its minimum energy? What is the minimum excitation energy from this state? What is the probability of finding it in the region of the molecule lying between $x = 0.49$ nm and $x = 0.51$ nm? What is the probability of finding it between $x = 0$ and $x = 0.2$ nm ?

Solution : Energies of a system are said to be quantized, that is, only some energies are possible because other energies correspond to untenable properties of the distribution of the particle. The Schrödinger equation accounts for those quantized energy levels. For a particle free to move in one dimension the Schrödinger equation reads

$$\left(\frac{-h^2}{2m}\right)\left(\frac{d^2}{dx^2}\right)\psi(x) + V(x)\psi(x) = E\psi(x)$$

where ψ is the wavefunction and V(x) is the potential energy of the particle, E is the total energy and $\hbar = h/2\pi$, a modification of Planck's constant. Consider an example of a particle of mass m which is free to move in a straight line between two walls separated by distance x.

The potential energy V (x) of the particle is constant and taken to be zero at all points and in the path of the particle except where it hits the wall. The Schrödinger equation for the region between x = 0 and x = L, where the potential energy is zero, is

$$\left(\frac{-\hbar^2}{2m}\right)\frac{d^2\psi}{dx^2} = E\psi(x) \qquad (1)$$

and its solution is in the form

$$\psi(x) = M\cos Kx + N\sin Kx \qquad (2)$$

where
$$K = \frac{\sqrt{2mE}}{\hbar} \qquad (3)$$

and M and N are constants. Since one end of the wall is at x = 0 the amplitude of the wave function using equation (2) is

$$\psi(x) = M\cos 0 + N\sin 0$$
$$= M$$

since sin 0 = 0 and cos 0 = 1.

But ψ (0) must be zero for the solution to be acceptable , thus M must be zero. This means that

$$\psi(x) = N \sin Kx.$$

At the other end of the wall, the amplitude is

$$\psi(L) = N \sin KL$$

which must be zero too. But ψ (0) = 0 for all x if N = 0 is not acceptable, thus KL must be chosen such that sin KL = 0. This implies that KL be some integral multiple of π since sin θ = 0 when θ = 0, π, 2π, ... Therefore the only values of K permitted are those for which

$$KL = n\pi \quad n = 1, 2, \ldots \qquad (4)$$

The relation between K and E is given in equation (3). Solving for E gives

$$K^2 = \frac{2mE}{\hbar^2}$$

and
$$E = \frac{K^2\hbar^2}{2m} \qquad (5)$$

Also from equation (4)

$$K = \frac{n\pi}{L}$$

Substituting for K in equation (5) shows that the energy of the system is confined to the values

$$E = \frac{n^2\pi^2\hbar^2}{2mL^2}$$

$$= \frac{n^2\pi^2h^2}{(2mL^2)(4\pi^2)}$$

$$= \frac{n^2h^2}{8mL^2} \qquad n = 1, 2, \ldots \qquad (6)$$

Thus the energy of the particle is quantized. The minimum energy is given as $E_2 - E_1$ where E_1 is the ground state (zero point) energy. Equat ion (6) would be used for calculating E_1 and E_2.

Therefore with n = 1,

$$E_1 = \frac{h^2}{8m_eL^2} = \frac{(6.626\times10^{-34}\,Js)^2}{8(9.110\times10^{-31}\,kg)(1\times10^{-9}\,m)^2}$$

where m_e = mass of electron particle

$= 6.024 \times 10^{-20}$ J $= 6.024 \times 10^{-23}$ kJ

$= 6.024 \times 10^{-23}$ kJ $\times$ 6.023 (10^{23}) mol^{-1}

$= 36.28$ kJ mol^{-1}

Using n = 2,

$$E_2 = \frac{2^2h^2}{8m_eL^2} = 4\times6.024\times10^{-20}\,J = 24.096\times10^{-20}\,J$$

$$= 4\times36.28\ kJ\ mol^{-1}$$

$$= 145.12\ kJ\ mol^{-1}$$

Thus $E_2 - E_1 = 145.12 - 36.28$

$= 108.\ 84$ kJ mol^{-1}

The distribution of the electron particle under consideration is given by

$$\psi_n(x) = \left(\frac{2}{L}\right)^{½} \sin\left(\frac{n\pi x}{L}\right) \qquad (2)$$

where n = 1.

For the first part, the region $0.49 \le x \le 0.51$ nm is almost infinitesimal and so the probability can be approximated by $[\psi_1^2 (x = 0.50)]\ \delta x$ with $\delta x = 0.51 - 0.49 = 0.02$ nm. From equation (2),

$$\psi_1^2(x=0.50) = \left(\frac{2}{L}\right)\sin^2\frac{\pi x}{L}$$

$$= \left(\frac{2}{1.0\times10^{-9}\,m}\right)\sin^2\left(\frac{0.50\pi}{1.0}\right)$$

$$= 2\times10^9\,m^{-1}$$

Therefore, the probability of being in the range 0.49 nm to 0.51 nm is

$$(2 \times 10^9\ m^{-1})\ (0.02\ nm) = 0.04.$$

For the second part, the region $0 \le x \le 0.2$ nm is not infinitesimal (on the scale of the molecule). Therefore, ψ_1^2 (x) dx has to be integrated numerically to get the total probability of being in that region.

The probability of being in the range x = 0 and x = 0.2 nm is therefore

$$\int_{0.0nm}^{0.2nm} \psi_1^2(x)dx = (2\times10^9\,m^{-1})\int_{0.0nm}^{0.2nm} \sin^2\left(\frac{\pi x}{1.0nm}\right)dx$$

$$= (2\times10^9\,m^{-1})\left[\frac{1}{2}x - \left(\frac{1.0nm}{4\pi}\right)\sin\left(\frac{2\pi x}{1.0nm}\right)\right]_{0.0nm}^{0.2nm}$$

$$= (2\times10^9\,m^{-1})\left[(0.1nm) - \left(\frac{1.0nm}{4\pi}\right)\sin(0.4\pi)\right]$$

$$= 0.0486$$

Q. 5 : The angular wave fuction of $\ell = 2$, m = 0 is given as

$$Y(\theta,\phi) = \left(\frac{5}{16\pi}\right)^{\frac{1}{2}}(3\cos^2(\theta) - 1)$$

and that for $\ell = 1$, m = 1 is

$$Y(\theta,\phi) = \left(\frac{3}{4\pi}\right)^{½} \sin\theta\cos\phi.$$

Show that these two functions are orthogonal.

Solution : Consider any dynamic system containing, say n particles. At any time the condition or state of such a system is given by the Schrodinger equation $\psi(q_1, q_2, ..., q_{3n}, t)$ where q_i represent the coordinates of the system, t is the time and ψ is the wave function. The physical signifiance of ψ is not clear but $\psi^*\psi$ can be interpreted as the probability of finding a particle within a certain coordinate in space. ψ^* is the complex conjugate of ψ. For a one dimensional problem the coordinate domain is the interval between q_i and $q_i + dq_i$, for a two-dimensional problem the domain is an element of areas, while the domain is an element of volume for a three dimensional problem.

If ψ is continuous and integrable, then

$$\int_{\text{all space}} \psi^* \psi \, dV = 1$$

where dV is the volume element and ψ is said to be normalized. On the other hand if

$$\int_{\text{all space}} \psi_i^* \psi_j \, dV = 0 \qquad (1)$$

for two wave functions ψ_i and ψ_j, then the two functions are said to be orthogonal.

In this problem there are two wave functions, so the integration can be carried out to see if their angular parts are zero. Then it can be concluded that the two functions are orthogonal

$$\psi^*\psi \, dV = (3\cos^2(\theta) - 1)(\sin\theta \cos\phi) \sin\theta d\theta d\phi$$

θ has values from 0 to 180° or π. ϕ takes values from 0 to 360° or 2π. Equation (1), using the limits for θ and ϕ, becomes

$$= \int_{\theta=0}^{\theta=\pi} \int_{\phi=0}^{\phi=2\pi} \left(\frac{5}{16\pi}\right)^{1/2} \left(\frac{3}{4\pi}\right)^{1/2} (3\cos^2(\theta) - 1)(\sin\theta\cos\phi)\sin\theta d\theta d\phi$$

$$= \int_{\theta}^{\pi} \int_{0}^{2\pi} \left(\frac{15}{64\pi_2}\right)^{\frac{1}{2}} (3\cos^2(\theta) - 1)(\sin\theta\cos\phi)\sin\theta d\theta d\phi$$

$$= \frac{\sqrt{15}}{8\pi} \int_0^{\pi} (3\cos^2(2\theta) - 1)\sin^2\theta d\theta \int_0^{2\pi} \cos\phi d\phi.$$

The integral over ϕ vanishes as follows.

$$\int_0^{2\pi} \cos\phi d\phi = \sin\phi\Big|_0^{2\pi}$$

$$= \sin 2\pi - \sin 0$$

$$= 0 - 0$$

$$= 0$$

$$\therefore \qquad \int \psi^* \psi = 0$$

Therefore, the two functions are orthogonal.

Q. 6 : The wave function, ψ, in spherical coordinates is written as

$$\psi (r, \theta, \phi) = R (r)\, \Theta (\theta)\, \Phi (\phi)$$

The radial contribution R is a function of r only and R^2 represents the probability of finding the electron at points r and r + dr in space for a hydrogen like atom. Given the values of r between 0 and $10a_0$, prepare plots of R(r) for the 1s, 2s, 3s, 3p and 3d subshells for hydrogen.

Solution : A hydrogen atom has a positive nucleus and one electron surrounding it. The Bohr radius a_0 is given as

$$a_0 = \frac{4\pi\varepsilon_0\hbar^2}{\mu e^2} \qquad (1)$$

where μ = the reduced mass for hydrogen. e is the electronic charge, ε_0 is the permittivity of vacuum and $\hbar = h / 2\pi$. h is Planck's constant. The reduced mass for hydrogen is given by

$$\mu = \frac{m_n m_e}{m_n + m_e}$$

where m_e = mass of the electron and m_n is the mass of the nucleus. Therefore,

$$\frac{1}{\mu} = \frac{m_n + m_e}{m_n m_e} = \frac{m_n}{m_n m_e} + \frac{m_e}{m_n m_e}$$

$$\frac{1}{\mu} = \frac{1}{m_e} + \frac{1}{m_n}$$

But $\frac{1}{m_n}$ is negligibly small compared to $\frac{1}{m_e}$. Therefore $\frac{1}{\mu} = \frac{1}{m_e}$ and $m = m_e$. Equation (1) becomes

$$a_0 = \frac{4\pi\left(8.85\times10^{-12}\frac{C^2s^2}{kg\,m^3}\right)\left(1.055\times10^{-34}\frac{kg\,m^2}{s}\right)^2}{(9.11\times10^{-3}kg)(1.602\times10^{-19}C)^2}$$

$$a_0 = (0.529\times10^{-10}m)\left(\frac{1Å}{10^{-10}m}\right)$$

or $\quad a_0 = 0.529$ Å.

Values of R(r) for different orbitals can be looked up in any physical chemistry text. For 1s orbital

$$R(r) = 2\left(\frac{Z}{a_0}\right)^{3/2} e^{-Zr/a_0}$$

For hydrogen atom Z = 1 thus for

1s orbital; $R(r) = 2\left(\frac{1}{a_0}\right)^{3/2} = e^{-r/a_0}$

2s orbital; $R(r) = \frac{1}{2\sqrt{6}}\left(\frac{1}{a_0}\right)^{3/2}\left(2-\frac{r}{a_0}\right)e^{-r/2a_0}$

2p orbital; $R(r) = \frac{1}{2\sqrt{6}}\left(\frac{1}{a_0}\right)^{3/2}\left(\frac{r}{a_0}\right)e^{-r/2a_0}$

3s orbital; $R(r) = \frac{1}{81\sqrt{3}}\left(\frac{1}{a_0}\right)^{3/2}\left(27-18\frac{r}{a_0}+2\left(\frac{r}{a_0}\right)^2\right)e^{-r/3a_0}$

3p orbital; $R(r) = \frac{4}{81\sqrt{6}}\left(\frac{1}{a_0}\right)^{3/2}\left(6-\frac{r}{a_0}\right)\frac{r}{a_0}e^{-r/3a_0}$

3d orbital; $R(r) = \frac{4}{81\sqrt{30}}\left(\frac{1}{a_0}\right)^{3/2}\left(\frac{r}{a_0}\right)^2 e^{-r/3a_0}$

Values of R(r) and $\frac{r}{a_0}$ for the different orbitals are plotted in the figure.

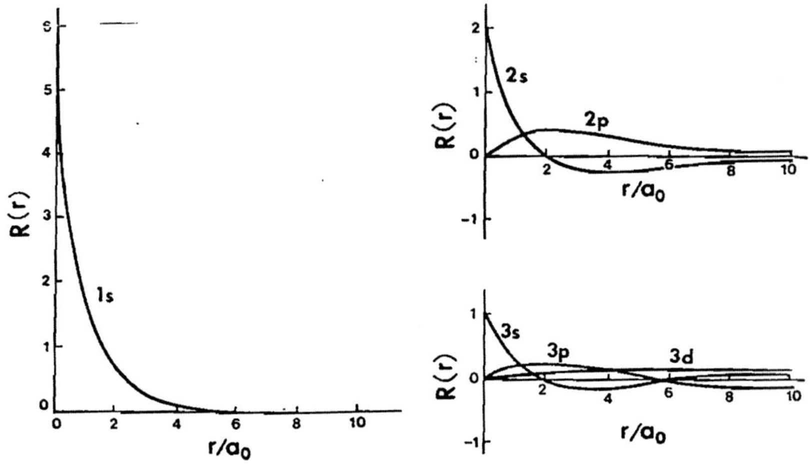

Q. 7 : Which of the following statements concerning the function of two of the four quantum number n and ℓ, with respect to their orbital, is/are correct.

(a) n determines the shape of an orbital

(b) ℓ determines the size of an orbital

(c) n determines the size of an orbital

(d) ℓ determines the shape of an orbital

(e) n determines the number of lobes of electron density in an orbital.

Solution : The solutions of the Schröndinger wave equation can be illustrated by means of electron probability diagrams called robtialls; six of which are shown in the figure. Looking at the first two columns of the figure, n increases but ℓ remains constant, and in the process, the size of an orbital type increases. But the shape and the number of lobes of the electron density remains unchanged. It can be said therefore, that n determines the size of orbitals whether they are spherical, dumbell or lobed, and not the shape. Statement (c) is thus correct while statements (a) and (e) are incorrect.

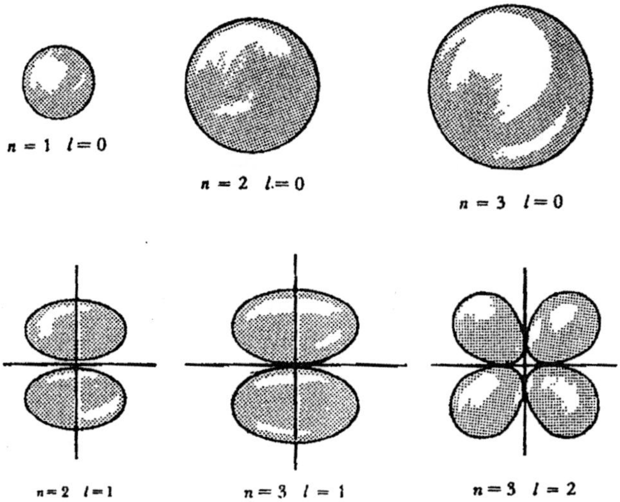

Fig.

Now as ℓ increases while n is kept constant, the shape of the orbitals change. The statement (d) is correct while statement (b) is incorrect. Statements (c) and a) are correct.

Q. 8 : Which of the following combinations of quantum numbers represent permissible solutions of the Schrödinger wave equation for the hydrogen atom :

	n	ℓ	m	s
(a)	3	0	1	$-\frac{1}{2}$
(b)	2	2	0	$+\frac{1}{2}$
(c)	4	3	–4	$-\frac{1}{2}$
(d)	5	2	2	$+\frac{1}{2}$
(e)	3	2	–2	$-\frac{3}{4}$

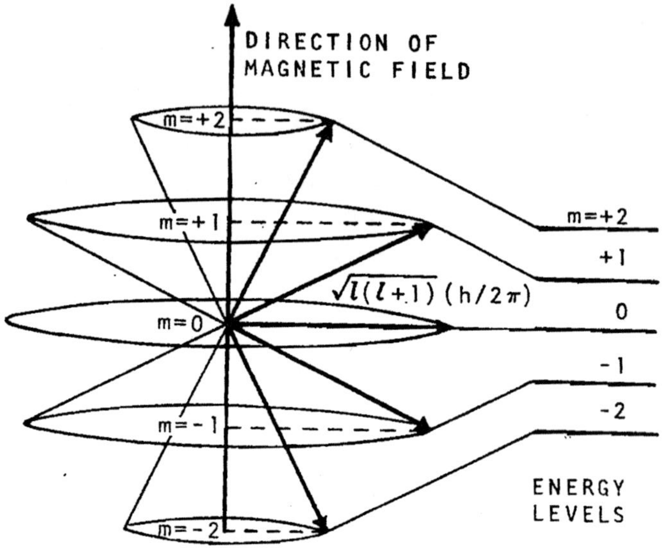

QUANTIZATION OF THE COMPONENTS OF ANGULAR MOMENTUM IN A MAGNETIC FIELD **B** FOR THE CASE $\ell=2$.

Fig.

Solution : The hydrogen atom is described by four quantum numbers: n, the principal quantum number, ℓ, the angular momentum quantum number, m, the magentic quantum number and s, the spin quantum number. Pauli exclusion principle states that, no two electrons in an atom can have the same four quantum numbers. Permissible values of the quantum numbers are :

$$n = 1, 2; 3, ...$$

$$\ell = 0, 1, 2, ..., (n-1)$$

$$m = -\ell, ..., 0, ..., +\ell$$

$$s = -\frac{1}{2} \text{ or } +\frac{1}{2}$$

In combination (a), $\ell = 0$ and m = 1. This is incorrect because according to permissible values for m, it cannot be greater than zero when $\ell = 0$. In combination (b), n = 2 and $\ell = 2$. This too is incorrect because when n = 2, $\ell = (n-1) = 1$, and as such ℓ cannot be greater than 1. $\ell = 3$ and $m = 4$ in combination (c); but m cannot be less than -3 when $\ell = 3$ since this will be contrary to the permissible values of m. Thus, statement (c) is incorrect. Combination (d) is correct since it is in accordance with the permissible values of the four quantum numbers. In combination (e) s is given as $-3/2$ but s cannot take any values other than $+1/2$ or $-1/2$ so combination (e) is incorrect.

Only combination (d) is correct and valid.

Q. 9 : At what value or values of r can nodes in the wave function for a 3s electron bound to a nucleus of charge +11 be predicted ? Bohr's radius = 0.0529 nm.

Solution : One must know two important things to answer this question : (1) The wave function expression for a one-electron atom for 3s, and (2) The fact that at a node, ψ or the wave function is zero. If ψ^2 is the probability of finding an electron in a given area and a node is an area of zero probability, then ψ must also be zero.

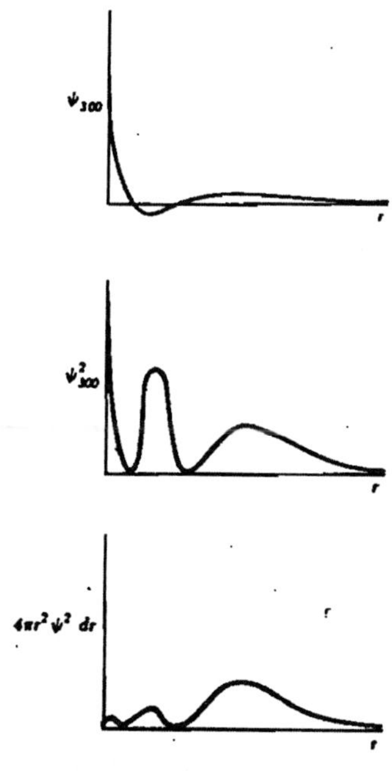

Graphs of ψ, ψ^2, and $4\pi r^2\psi^2\,dr$ for a 3s electron.

The wave function expression relates ψ to radius r.
Then wave function for a 3s electron is

$$\psi_{300} = \frac{1}{81\sqrt{\pi}}\left(\frac{Z}{a_0}\right)^{3/2}\left(27 - \frac{18Zr}{a_0} + \frac{2Z^2r^2}{a_0^2}\right)e^{-Zr/3a_0}.$$

For ψ_{300} to be equal to zero the factor $\left(27 - \frac{18Zr}{a_0} + \frac{2Z^2r^2}{a_0^2}\right)$ must be equal to zero because the other factors in the equation can never be equal to zero.

Using $O = 27 - \frac{18Zr}{a_0} + \frac{2Z^2r^2}{a_0^2}$ solve for r.

$$O = 27 - \frac{18(11)(r)}{(.0529)} + \frac{2(11)^2(r)^2}{(.0529)^2}$$

$$= 27 - 3.74 \times 10^3 r + 8.67 \times 10^4 r^2$$

Using the quadratic equation one finds that r is equal to 0.0092 or 0.034 nm.

Q. 10 : The mass of an electron is 9.11×10^{-31} and Planck's constant = 6.63×10^{-34} J sec. How much energy does an electron possess in its five lowest states, if it is placed in a cube of dimension 50 nm ?

Solution : Wave function theory states that the permitted values of the energy of a particle in a box is given as

$$E = \frac{N^2h^2}{8mL^2},$$

where E = energy, N = principal quantum number, h = Planck's constant, m = mass of particle, and L = length of box. Substitute in the known values and solve for E. The principal quantum number N corresponds to the energy levels of the electrons. The five lowest energy states are N = 1, N = 2, N = 3, N = 4, and N = 5, where N = 1 is the lowest and N = 5 the highest energy level.

Substituting:

$$E = \frac{N^2h^2}{8mL^2} = \frac{N^2(6.63\times10^{-34}\,J\,sec)^2}{8(9.11\times10^{-31}\,kg)(5\times10^{-8}\,M)^2}$$

$$= N^2(2.41\times10^{-23})$$

(Note : 50 nm = 5.0×10^{-8} M)

1st energy level (lowest)	:	$N^2 = 1$, $E = 2.41 \times 10^{-23}$ J
2nd energy level	:	$N^2 = 4$, $E = 9.64 \times 10^{-23}$ J
3rd energy level	:	$N^2 = 9$, $E = 2.16 \times 10^{-22}$ J
4th energy level	:	$N^2 = 16$, $E = 3.85 \times 10^{-22}$ J
5th energy level	:	$N^2 = 25$, $E = 6.025 \times 10^{-22}$ J.

Q. 11 : Give qualitative explanation for the observed splitting of the ground state energy of the hydrogen atom in the presence of a magnetic field.

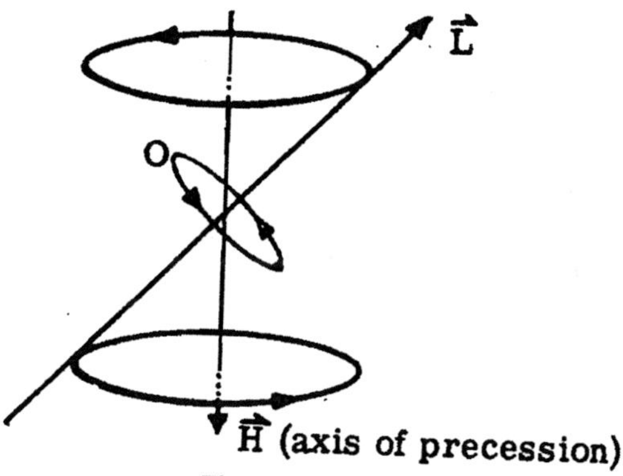

Fig.

Solution : This phenomenon is called Zeeman splitting, after the Dutch physicist Pieter Zeeman who first discovered the effect in 1896.

Consider an electron moving in the ground state orbital of a hydrogen atom and describing a circular orbit (labelled "O" in the diagram). This motion will give rise to an angular momentum vector $\vec{L}$ directed along the axis of rotation. In the presence of a magnetic field $\vec{H}$, the angular momentum vector will precess about the magnetic field axis.

Now, since the angular momentum of an electron must be quantized, the projection of $\vec{L}$ onto $\vec{H}$ (that is, the component of the vector $\vec{L}$ along axis $\vec{H}$) must also be quantized. From quntum theory, we know that this projection must take on values which are integral multiples (positive, negative, or zero) of $h/2\pi$, where h is Planck's constant. These integers are the possible values of the magnetic quantum number, m. For the hydrogen atom, the magentic quantum number can take on the values m = –1, 0, +1.

Thus, we observer three lines for hydrogen, in a magnetic field. The high energy line corresponds to the case where the projection of $\vec{L}$ onto $\vec{H}$ has the same direction as $\vec{H}$ (m = 1), the intermediate line occurs when $\vec{L}$ and $\vec{H}$ are perpendicular and the projection of $\vec{L}$ onto $\vec{H}$ is thus zero (m = 0), and the low energy line corresponds to the case where the projection of $\vec{L}$ onto $\vec{H}$ has the opposite direction of $\vec{H}$ (m = –1).

5

Chemical Bonding

Q. 1 : H_2O_3, hydrogen trioxide, a close relative of hydrogen peroxide, has recently been synthesized. It is extremely unstable and can be isolated only in very small quantities. Write a Lewis electron dot structure for H_2O_3.

Solution : A Lewis dot structure of a compound shows the arrangment of valence electrons. Valence electrons are defined as an element's outer electrons which participate in chemical bonding. Thus, to write an electron dot structure for H_2O, calculate the total number of valence electrons.

This can be done by considering the electronic configurations of H (hydrogen) and 0 (oxygen).

Hydrogen possesses one valence electron while oxygen has 6 valence electrons. In H_2O_3, a total of $(2)1 + (3)6 = 20$ valence electrons are involved.

With this in mind, the Lewis electrons dot structure becomes

```
    ..    ..    ..
H : O  :  O :   O :H
    ..    ..    ..
```

Notice, you have represented the required 20 valence electrons.

Q. 2 : Using electron-dot notation, show for each of the following the **outer** shell electrons for the uncombined atoms and for the molecules or ions that result:

(a) H + H → hydrogen molecule

(b) Br + Br → bromine molecule

(c)	Br	+	Cl	→	bromine chloride
(d)	Si	+	F	→	silicon fluoride
(e)	Se	+	H	→	hydrogen selenide
(f)	Ca	+	O	→	calcium oxide

Solution : When electrons are transferred from one atom to another, ions are formed, which gives rise to ionic bondign. Two atoms, both of which tend to gain electrons, may combine with each other by sharing one or more pairs of electrons. These two atoms form a covalent bond.

To solve this problem, one must know the number of valence electrons, in each of the atoms in the equations. The valence number reflects the combining capacity of an atom. Next, one must know which atoms combine to form ionic bonds and which form covalent bonds. The only ionic bond formed in these equations is for Ca + O; the other bonds are covalent, and electrons are shared to form an isoelectronic electron cloud such as a noble gas.

Thus

(a) H˙ + H˙ → H:H

(b) :B̈: + :B̈: → :B̈: :B̈:

(c) :B̈: + :C̈l: → :B̈: :C̈l:

(d) ·Si· + :F̈: → :F : S: F: (with :F: above and :F: below S)

(e) ·S̈e: + H˙ → H : S̈e: with Ḧ below

(f) Ca· + ·Ö: → Ca^{++} : Ö $^{=}$

Q. 3 : The sulfate ion consists of a central sulfur atom with four equivalent oxygen atoms in a tetrahedral arrangement. Keeping in mind the octet rule, draw the electronic structure for the ion. What should the internal O–S–O bond angle be ?

Solution : The formula for the sulfate ion is $SO_4^{=}$. To write the electronic structure of the ion in this problem, one must consider the definition of valence and the octet rule. Valence electrons refer to those outer electrons that participate in chemical bonding. The octet rule states that for stability, there can be no more than eight electrons in the outer orbit of an atom, either as a result of transfer or sharing. Thus, after determining the number of valence electrons present, arrange them so that they obey the octet rule. The electronic configuration of sulfur is $1s^2 2s^2 2p^6 3s^2 3p^4$. It has 6 valence electrons; they are found in the 3s and 3p orbitals. The electronic configuration of oxygen is $1s^2 2s^2 2p^4$, which means it also has six valence electrons; the first orbital contains only 2 electrons. The electronic structure of each of these atoms can be represented as shown in figure a and figure b, where x's and dots indicate valence electrons. The electronic structure of the ion can be pictured as figure. c. Here, one can see that the sulfur and two of the oxygen atoms are surrounded by eight electrons. Because two of the oxygen atoms are only surrounded by seven electrons, each of these atoms possesses a negative charge. This structure may also be drawn as shown in figure d.

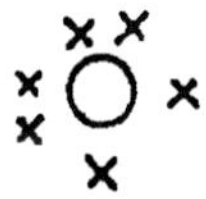

Fig. a

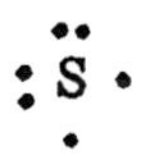

Fig. b

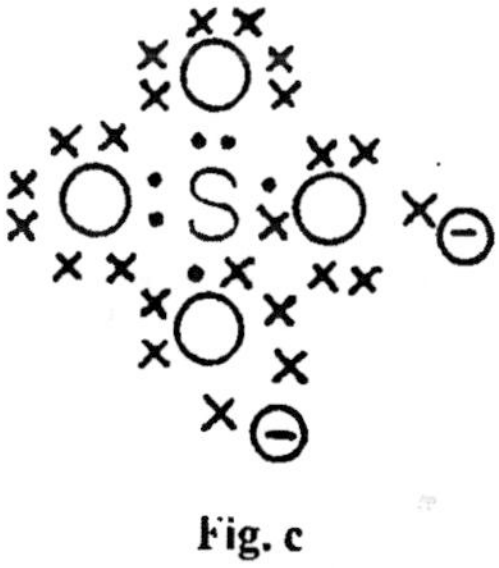

Fig. c

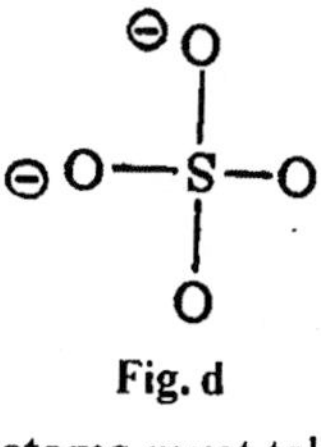

Fig. d

For greatest stability the oxygen atoms must take positions as far apart as possible. It has been shown that when four atoms surround a fifth the most stable arrangement is tetrahedral. In this arrangement, the O–S–O bonds form a 109.5° angle.

Q. 4 : Distinguish a metallic bond from an ionic bond and from a covalent bond.

Solution : The best way to distinguish between these bonds is to define each and provide an illustrative example of each.

When an actual transfer of electrons results in the formation of a bond, it can be said that an ionic bond is present. For example,

$2K^{\circ}$	+	$\ddot{\underset{..}{S}}:$ $\rightarrow$	$2K^{+}$	+	$\ddot{\underset{..}{S}}:^{2-}$	+	K_2S
potassium atoms		sulfur atom	potassium ions (unlike ions due to transfer of electrons from potassium to sulfur)		sulfur ion		ionic bond due to the attraction of unlike ions

When a chemical bond is the result of the sharing of electrons, a covalent bond is present. For example :

$$:\ddot{\underset{..}{Br}}\cdot + \cdot\ddot{\underset{..}{F}}: \rightarrow :\ddot{\underset{..}{Br}}:\ddot{\underset{..}{F}}:$$

These electrons are shared by both atom.

A pure crystal of elemental metal consists of millions of atoms held together by metallic bonds. Metals possess electrons that can easily ionize, i.e., they can be easily freed from the individual metal atoms. This free state of electrons in metals binds all the atoms together in a crystal. The free electrons extend over all the atoms in the crystal and the bonds formed between the electrons and positive nucleus are electrostatic in nature. The electrons can be pictured as a "cloud" that surrounds and engulfs the metal atoms.

Q. 5 : An ionic bond is established between positive ion A and negative ion B. How would one expect the strength of the bond to be affected by each of the following changes : (a) Doubling the charge on A, (b) Simultaneously doubling the charge on A and B, (c) Doubling the radius of B, and (d) Simultaneously doubling the radius of and B ?

Solution : When two ions come together to form an ionic bond, energy is released. The more energy released, the stronger the ionic bond. The amount of energy released is determined by the equation

$$E = 1.44\frac{q_1 q_2}{r^2}$$

where E = energy in electron volts, q_1 = charge on the positive ion, q_2 = charge on the negative ion, and r = the distance between the nuclear centers. If one changes one parameter, another must adjust to maintain this equality.

Solving :

(a) If the charge on A, the positive ion, is doubled, the energy released is doubled. That is, $2E = 1.44\frac{2q_1q_2}{r^2}$ is equivalent to $E = \frac{1.44q_1q_2}{r^2}$.

If E is doubled, the strength of the chemical bond is doubled.

(b) Using a similar line of reasoning, it is found that if one doubles both the positive and negative charges, four time E is released. The strength of the bond is, thus, 4 times as great as the original bond.

(c) If one increeases the radius of the ions then, the distance between their nuclei must also increase. To maintain the equality. E must decrease because the square of r is inversely proportional to E. Therefore, the bond strength decreases.

(d) Both radii are doubled. E is diminished by a factor of 4 and the strength of the bond is decreased by a factor of 4.

Q. 6 : Compare the ionic bond strength of NaCl to KF; assuming that the radii of Na^+, K^+, F^-, and Cl^- are, respectively, 0.097, 0.133, 0.133, and 0.181 NM.

Solution : The ionic bond strength is directly proportional to the energy released when positive and negative ions form chemical bonds. To solve this problem, therefore, compute how much energy is released when Na^+ and Cl^- ions form NaCl and when K^+ and F^- ions form KF. Employ the expression that relates energy to charge and radii :

$$E = 1.44\frac{q_1q_2}{r^2}$$

where E = energy in electron volts, r = distance between the nuclei of the ions in the molecule, and q_1 and q_2 are the charges on the ions. Note that the charge on each ion is the same. The distance between the nuclei, r, is the sum of the radii of the ions. With these facts in mind, one can determine the amount of energy released in the formation of NaCl.

$$E_{NaCl} = 1.44\frac{q_1q_2}{0.0773}. \text{ For KF, } E_{KF} = 1.44\frac{q_1q_2}{0.0708}$$

Comparing the amount of energy released in each case

$$\frac{E_{NaCl}}{E_{KF}} = \frac{0.0708}{0.0773}$$

$$\frac{0.266}{0.278} \times 100 = 91.6\%.$$

Thus, the bond strength of NaCl is 91.6% as strong as the bond strength of KF.

Q. 7 : A chemist possesses KCl, PH_3, $GeCl_4$, H_2S, and CsF. Which of these compounds do not contain a covalent bond ?

Solution : A covalent bond is defined as one in which electrons are shared. The stability of covalent bonds in molecules depends on the difference in electronegativity values of the two atoms which make up the molecule. Electronegativity refers to the tendency of an atom to attract shared electrons in a chemical bond. If the electronegativity difference of two elements is greater than 1.7, an ionic bond is formed; if it is less than 1.7, a covalent bond is formed.

To solve this problem, consult a table of electronegativity values, and compute the electronegativity difference of the atoms in each of the given compounds. Proceed as follows :

	Electronegativity Values		**Difference**
KCl	K = 0.8	Cl = 3.0	2.2
PH_3	P = 2.1	H = 2.1	O
$GeCl_4$	Ge = 1.8	Cl = 3.0	1.2
H_2S	H = 2.1	S = 2.5	0.4
CsF	Cs = 0.7	F = 4.0	3.3

Thus, only KCl and CsF exceed 1.7. They possess ionic bonds. The remainder of the molecules possess covalent bonds.

Q. 8 : Consider a covalent bond between hydrogen and arsenic. It is known that the radii of hydrogen and arsenic atoms are respectively: 0.37 and 1.21 Angstroms. What is the approximate length of the hydrogen-arsenic bond ?

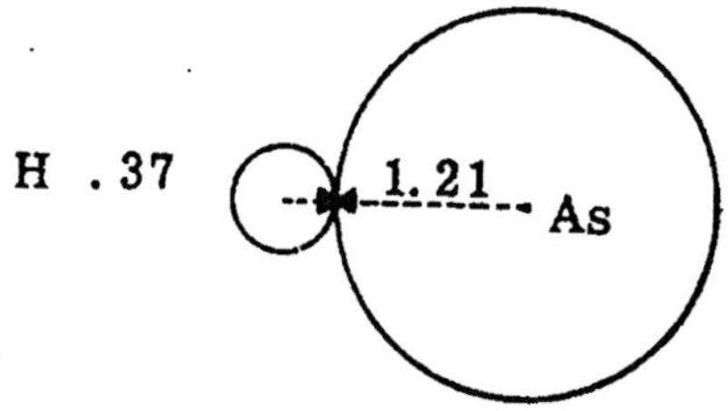

Fig.

Solution : One may assume that the two atoms are spheres. Thus, as in the figure shown, the bond length defined as the distance between the two nuclei is the sum of the lengths of the two radii.

bond length = length of H radii + length of As radii.

bond length = 0.37Å + 1.21 Å = 1.58 Å.

Q. 9 : Hydrogen chloride (HCl) is used industrially in the manufacture of fabric dyes. The acidic properties of HCl arise because the H–Cl bond is polar. The observed dipole moment, u_{obs}, of HCl is 1.03 D and the H–Cl bond distance is 1.28 Å. Estimate the percent ionic character of the HCl bond. Electron charge, e = 4.8×10^{-18} esu.

Solution : A dipole moment is an indication of bond polarity. The magnitude of the dipole moment is the product of the electronic charge and the distance between the charge centers.

To determine the ionic character of HCl, the magnitude of the dipole moment is needed.

Percent ionic character $= \frac{u_{obs}}{u_{ionic}} \times 100$, where u_{ionic} = the magnitude of dipole moment for total ionic character. If HCl were ionic, one electron would be transferred from H to Cl to give te ionic structure H+Cl–.Thus, the electron charge would be the charge on one electron, 4.8×10^{-18} esu. The distance between the two atoms is given as 1.28 Å or 1.28×10^{-8} cm. (There are 10^{-8} cm per Å). Recalling the definition of dipole moment,

$$u_{ionic} = 4.80 \times 10^{-10} \text{ esu} \times 1.28 \times 10^{-8} \text{ cm}$$
$$= 6.14 \times 10^{-18} \text{ esu–cm.}$$

Dipole moment is usually expressed, however, in Debyes. There are 10^{18} Debyes per esu–cm. Thus,

$$6.14 \times 10^{-18} \text{ esu–cm} = 6.14 \text{ Debyes} = u_{ionic}$$

Therefore, percent ionic character

$$= \frac{1.03}{6.14} \times 100\% = 16.8\%.$$

Q. 10 : Describe the bonding in linear, covalen $BeCl_2$ and planar, covalent BCl_3. What is the difference in the hybrid orbitals used ?

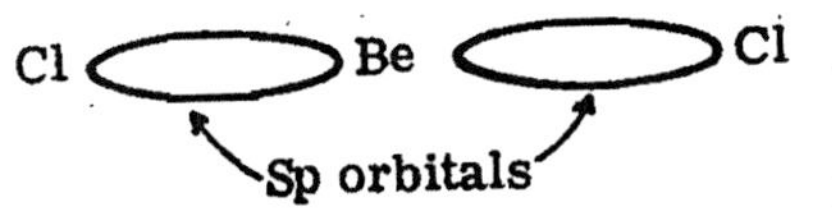

Fig. A

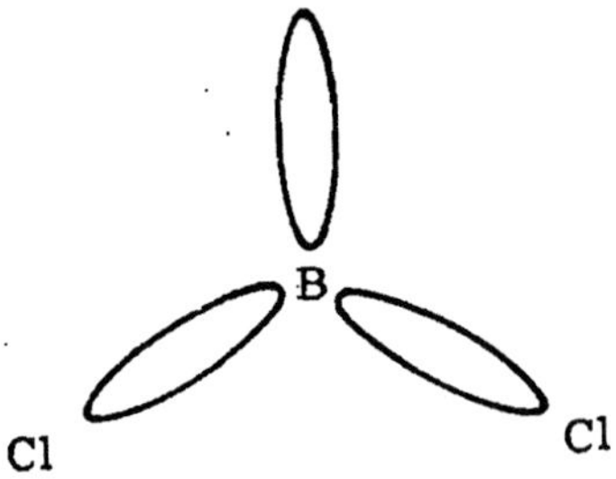

Fig. B

Solution : The solution to this problem involves the hybridization of orbitals. Once this is clear, the bonding in $BeCl_2$ and BCl_3 will follow.

Quantum theory deals with independent orbitals, such as 2s and 2p. This can be applied to a species, like hydrogen, with only 1 electron. However, with atoms that contain more than one electron, different methods must be used. For example, the presence of a 2s electron perturbs a 2p electron, and vice-versa, such that a 2s electron makes a 2p electron take on some s-like characterisitics. The result is that the hydrogen like 2s and 2p orbitals are replaced by new orbitals, that contain the combined characteristics of the original orbitals. These new orbitals are called hybrid orbitals. The number of hybrid orbitals resulting from hybridization equals the number of orbitals being mixed together. For example, if one mixes an s and p orbital, one obtains two sp hybrid orbitals. One s and two p = three sp^2 orbitals. One s with three p = four sp^3 orbitals. sp orbitals are linear. sp^2 orbitals assume a planar shape and sp^3 orbitals assume a tetrahedral shape.

Solving : It is given that $BeCl_2$ is linear and covalent. Since sp orbitals are linear, Be undergoes sp hybridization. If something is linear, the bond angle is 180°. By understanding hybridization, one also knows the geometry of the molecule. A diagram of the bonding resembles Fig. A.

Given that BCl_3 is planar, and since sp^2 hybridization yields a planar structure, B has sp^2 hybridized bonding with angles of 120° (Fig. B).

Q. 11 : Draw Lewis structures with all appropriate resonance forms, for each of the following : (a) SO_2 ; (b) NO_2^- ;

Solution : There are molecules in which the arrangement of atoms is not adequately described by a single Lewis structure. Therefore, all the possible Lewis structures are used to represent them. This concept is called resonance. It does not imply that the two or more forms are different or that the molecule really exists in different forms. There is only one form of the molecule. The only differences are in how the arrangements of electrons are represented.

(a) SO_2,

These structures satisfy the octet rule because each atom is surrounded by eight electrons. A bond represents two electrons.

(b) NO_2^-

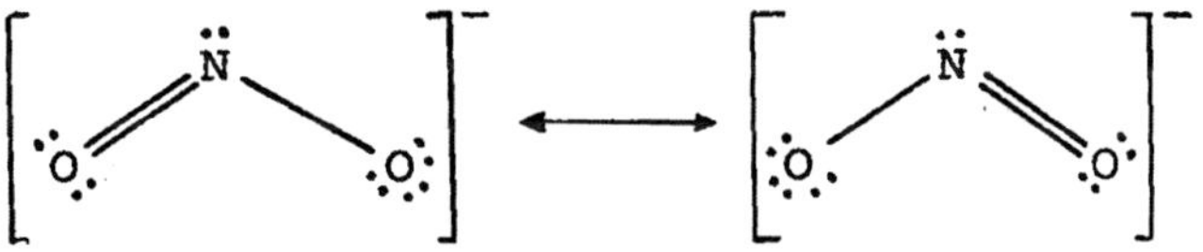

The double-headed arrow is used to indicate that both structures are resonance forms. Note that the arrangement of the nuclei is the same in both structures differing only in the placement of the electrons.

Q. 12 : Assuming that the bond angles around the central carbon are tetrahedral in all cases, arrange the following compounds in the expected order of increasing polarity: (a) CF_3Cl; (b) CF_4; (c) CF_2Cl_2; (d) CF_2H_2; (e) CH_4.

Solution : Polarity results when one atom in a compound exerts greater attraction for a shared electron pair than the other atom, causing the shared electron pair to be displaced to the one atom. A measure of polarity is obtained from the electronegativity difference between the atoms in the compound. The higher the difference between any two atoms, the polar the bond becomes. The halogens on the far right hand part of the periodic table have very high electronegativities so that if any of halogrns form bonds with atoms to the left or middle of the table then we can expect a large difference in electronegativities.

(a)

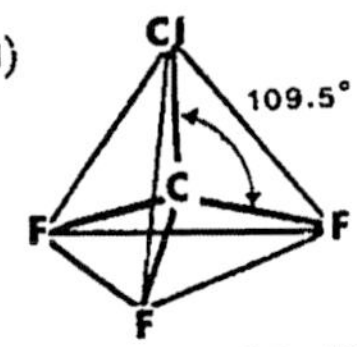

CF_3Cl

(b)

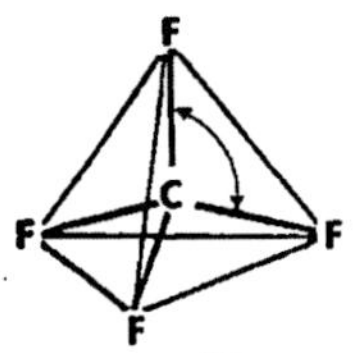

CF_4

(c)

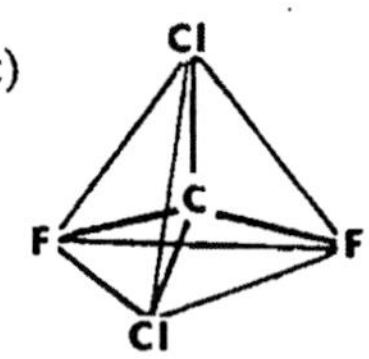

CF_2Cl_2

(d)

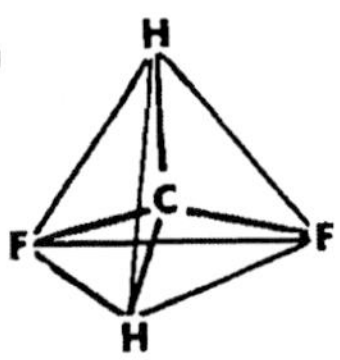

CF_2H_2

(e)

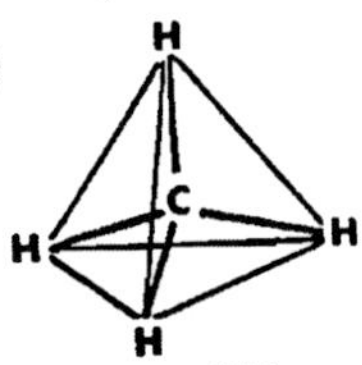

CH_4

In compound (a) three flourine atoms which are highly electronegative and one chlorine atom which is also very electronegative form bonds with one carbon atom. One would expect then that the shared electron pairs of the compound would be somewhat shared equally between the two electronegative atoms. In compounds (c) and (d) the central carbon atoms form bonds with equal number of Cl_2 and F_2; and H_2 and F_2, respectively. Compound (c) should be more polar because the electronegativity difference between the F_2 and H_2 bond is greater than that between the F_2 and Cl_2 bond. If the bond angles around the central carbon are tetrahedral, then the structures of the compounds are as follows :

Equal sharing of a pair of electrons occurs between atoms which are identical. Therefore there is no polarity between such atoms. However, it is possible for a molecule to contain bonds which are polar and yet still be nonpolar itself. This comes about because of the geometry of the molecule. The overall polarity of the molecule is found by combining the polarity of the indvidual bonds of the molecule vectorially. In certain molecules, i.e. CH_4, CF_4, the geometry allows the vectors to cancel, producing a nonpolar molecule.

Q. 13 : When potassium metal (K) is burned in dry air or dry oxgen, bright yellow potassium superoxide (KO_2) is formed. What is the ground state electronic configuration of the superoxide ion (O_2^-) in potassium superoxide on the basis of molecular orbital thoery ?

Solution : In the same way that electrons in atoms exist in allowed energy states called atomic orbitals, electrons exist in molecles in allowed energy states called molecular orbitals. An arrangement of these electrons can be seen by considering a simple diatomic molecule of hydrogen. The electronic configuration of H_2 is $1s^2$.

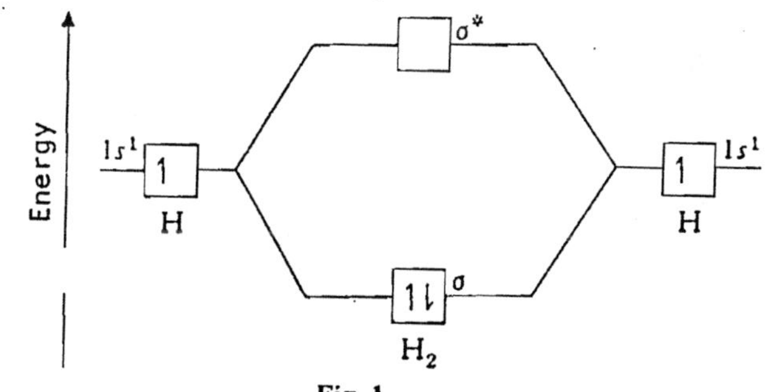

Fig. 1

From fig. 1, one 1s orbital of a hydrogen atom combines with one 1s orbitals of the other hydrogen atom to form two molecular orbitals. One of the combination leads to the bonding orbital called σ, which is a lowering of energy. The other orbital leads to instability (due to a raising of energy) and is called the antibonding orbital σ*.

Now consider the formation of Li_2 from two Li atoms. The

electronic configuration of the Li atom is $1s^2 2s^1$. When two Li atoms combine, there could be a possible interaction between the 1s orbital of one Li atam and the 2s orbital of the other atom. But such an interaction is of no importance since atomic orbtials combine most effectively with other atomic orbtials of similar energy. The energy difference between the 1s and 2s orbitals is too great for any possible interaction, thus the energy-level diagram for any possible interaction, thus the energy-level diagram for Li_2 molecule is as shown in Fig. 2. Only the 2s electrons contribute to bonding in the Li_2 molecule. The molecular orbitals are labelled with subscripts to indicate the set of atomic orbitals from which they are formed.

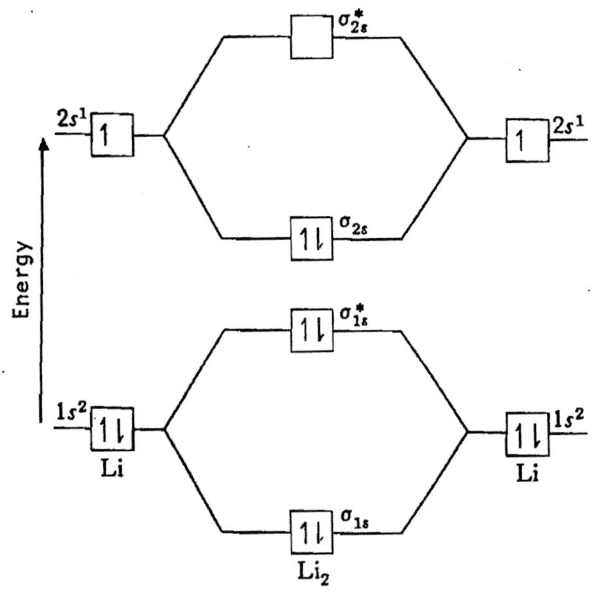

Fig. 2

Whenever an atom has a completed s, p, or d level, the electrons in those orbitals do not contribute to bonding. If the electronic configuration is written to include the p atomic orbtials, combination of the s and p orbitals can lead to nonzero overlap. When the p orbitals overlap in bonding, they change sign about the bond axis. The molecular orbitals thus formed are labeled π.

The bonding in the diatomic molecule of O_2^- can now be described. O_2^- is an A_2 type molecule. There are 17 electrons in O_2^- (eight from each of the two O atoms and one from the net charge). Four of these are not valence electrons and are not involved in bonding. Therefore, there are 13 electrons to feed into our molecular orbitals. The resulting ground state electronic configuration is

$$KK\ (\sigma_s)^2\,(\sigma^*_s)^2\,(\pi_{x,y})^4\,(\sigma_z)^2\,(\pi_x^*)^2\,(\pi^*_y)^1$$

Note: $(\pi_x^*)^2\,(\pi^*_y)^1$ could also be written as $(\pi_{x,y}^*)^3$ and that the symbol KK represents the four inner electrons (two from each O atoms) that are not involved in bonding.

Q. 14 : The ion Ti^{3+} has one unpaired electron. Calculate the contribution of the spin of this electron (a) to the magnetic moment of the ion and (b) to the molar magnetic susceptibility of Ti^{3+} at 298 K.

Solution : (a) The quantum number, s of Ti^{3+} is equal to $\frac{1}{2}$. The contribution of the spin of this electron to the magnetic moment of the ion is given by

$$\begin{aligned} \mu &= g\sqrt{s(s+1)}\hbar \\ &= g\sqrt{½(½+1)}\hbar \\ &= g\tfrac{\hbar}{2}\sqrt{3} \\ &= \sqrt{3}\tfrac{gh}{4\pi} = \sqrt{3}\beta \end{aligned} \qquad (1)$$

Note that $\hbar = h/2\pi$ and h = Planck's constant. $\beta = gh/4\pi$ is known as the Bohr magneton. Therefore,

$$\begin{aligned} \mu &= \sqrt{3}(0.927\times10^{-20}\text{erg. gauss}^{-1}) \\ &= 1.606\times10^{-20}\text{erg gauss}^{-1} \end{aligned}$$

(b) The contribution of the spin of this electron to the molar magnetic susceptibility of Ti^{3+} at 298° K is given by

$$x_m = \frac{4\beta^2 Ls(s+1)}{3kT} \qquad (2)$$

where L = Avogadro's number = 6.02×10^{23}
k = constant = 1.3805×10^{-16}
T = temperature = 298° K

Since all the variable in equation (2) are known, the problem becomes a matter of mere substitution. Hence,

$$x_m = \frac{4(0.927\times10^{-20})^2(6.02\times10^{23})½(½+1)}{3(1.3805\times10^{-16})(298)}$$

$$\therefore \quad x_m = 0.01258 \times 10^{-1}$$
$$= 1.258 \times 10^{-3}$$

Q. 15 : The density of $SiHBr_3$ is 2.690 $g.cm^{-3}$ at 25°C, its refractive idex is 1.578, and its dielectric constant is 3.570. Estimate its dipole moment in debyes.

Solution : The molecular weight of $SiHBr_3$ = 28 + 1 + 3 (80) = 29 + 240 = 269. The molar polarization of a substance is given by

$$P_m = \frac{\varepsilon/\varepsilon_0 - 1}{\varepsilon/\varepsilon_0 + 2}\frac{M}{d} \qquad (1)$$

where $\varepsilon/\varepsilon_0$ = dielectric constant

M = molecular weight

d = density.

From equation (1),

$$P_m = \left(\frac{3.570-1}{3.570+1}\right)\frac{269\text{g mole}^{-1}}{2.690\text{g cm}^{-3}}$$

$$= \left(\frac{2.570}{5.570}\right)\frac{269\text{g mole}^{-1}}{2.690\text{g cm}^{-3}}$$

$$\therefore \quad P_m = 46.14 \text{ cm}^3 \text{ mole}^{-1} \qquad (2)$$

The molar refractivity R_m, is defined as

$$\left(\frac{n^2-1}{n^2+2}\right)\frac{M}{d}$$

where n = refractive index. Therefore

$$R_m = \left(\frac{(1.578)^2-1}{(1.578)^2+2}\right)\frac{269\text{g mole}^{-1}}{2.69\text{g cm}^{-3}}$$

$$= \left(\frac{1.49}{4.49}\right)100 \text{ cm}^3 \text{ mole}^{-1}$$

$$\therefore \quad R_m = 33.184 \text{ cm}^3 \text{ mole}^{-1} \qquad (3)$$

The difference between the molar polarization and molar

refractivity is related to dipole moment by

$$P_m - R_m = \frac{4\pi L}{3}\left(\frac{\mu^2}{3kT}\right) \quad (4)$$

where L = Avogadro's constant = 6.02×10^{23} mole^{-1}
μ = dipole moment
k = Boltzmann's constant = 1.38051×10^{-16} erg k^{-1}
T = Absolute temperature

Using equations (2) and (3) the difference is

$$P_m - R_m = (46.14 - 33.184)\ \text{cm}^3\ \text{mole}^{-1}$$
$$= 12.956\ \text{cm}^3\ \text{mole}^{-1}$$

substitute $P_m - R_m$ into equation (4)

$$12.956 = \frac{4\pi L}{3}\left(\frac{\mu^2}{3kT}\right) \quad (5)$$

Rearranging equation (5) to solve for μ yields

$$\mu = \sqrt{\frac{(3)(3kT)(12.956)\text{cm}^3\text{mol}^{-1}}{4\pi L}}$$

$$= \sqrt{\frac{9(1.38051\times10^{-16})\text{erg}^\circ\text{K}^{-1}(298^\circ\text{K})(12.956)\text{cm}^3\text{mole}^{-1}}{4(3.142)(6.02\times10^{23})\text{mole}^{-1}}}$$

$$= \sqrt{\frac{9(1.3805\times10^{-16})(298)(12.956)\text{erg}^\circ\text{K}^{-1}\,{}^\circ\text{Kcm}^3\text{mole}^{-1}}{4(3.142)(6.02\times10^{23})\qquad\qquad \text{mole}^{-1}}}$$

$$= 7.97\times10^{-19}\sqrt{\text{cm}^3\text{erg}}$$

$$= 7.97\times10^{-19}\sqrt{\text{cm}^3\text{g cm}^2\ \text{sec}^{-2}}$$

$$= 7.97\times10^{-19}\sqrt{\text{g cm sec}^{-2}\text{cm}^4}$$

$$= 7.97\times10^{-19}\sqrt{\text{dyne cm}^2\text{cm}^2}$$

Since $1\ (\text{esu})^2 = \text{dyne cm}^2$,

$$= 7.97 \times 10^{-19}\ \text{esu.cm}$$

$\therefore$ μ =0.7972 D

Q. 16 : The dielectric constant of gaseous SO_2 is 1.0093 at 273 K and 1.00569 at 373 K, at P = 1 atm. Calculate the dipole moment of SO_2. Assume ideal gas behavior.

Solution : The molar polarization of a substance is related to the dielectric constant by

$$P_m = \frac{\varepsilon/\varepsilon_0 - 1}{\varepsilon/\varepsilon_0 + 2}\frac{M}{d} \tag{1}$$

where P = molar polarization

$\varepsilon/\varepsilon_0$ = dielectric constant

Considering the polarization of the molecule to result from permanent and induced dipole moments, equation (1) changes to the form

$$P_m = \frac{4\pi L}{3}\left(\alpha + \frac{\mu^2}{3kT}\right) \tag{2}$$

where α = ploarizability of the molecule

μ = permanent dipole moment

and L = Avogadro's number

A plot of P against $\frac{1}{T}$ will have a slope of

$$\frac{4\pi L\mu^2}{9k}$$

$$\frac{M}{d} = \frac{RT}{P} = 82.06\ \text{cm}^3\ {}^\circ\text{K}^{-1}\ \text{mole}^{-1}(T).$$

Therefore,

$$P_m(\text{at } 273^\circ\text{K}) = \left(\frac{1.00993-1}{1.00993+2}\right)(82.06)\text{cm}^3\ \text{K}^{-1}\ \text{mole}^{-1}(273)^\circ\text{K}$$

$$= 73.99\ \text{cm}^3\ \text{mole}^{-1}$$

and

$$P_m(\text{at } 373^\circ\text{K}) = \left(\frac{1.00569-1}{1.00569+2}\right)(82.06)\text{cm}^3\ {}^\circ\text{K}^{-1}\ \text{mole}^{-1}(373)^\circ\text{K}$$

$$= 58.10\ \text{cm}^3\ \text{mole}^{-1}$$

from these,

$$\text{slope} = \frac{P_m(\text{at } 373^\circ\text{K}) - P_m(\text{at } 273^\circ\text{K})}{\frac{1}{373} - \frac{1}{273}}$$

$$= \frac{(58.10 - 73.99)\text{cm}^3 \text{ mole}}{(0.00268 - 0.00366)\frac{1}{^\circ\text{K}}}$$

$$= 16.20 \times 10^3 \text{ cm}^3 \ ^\circ\text{K mole}^{-1}$$

The constant $\frac{4\pi L}{9k} = 6.090 \times 10^{39} \text{erg}^{-1} \ ^\circ\text{K mole}^{-1}$. Therefore,

$$\mu^2 = \frac{16.20 \times 10^3 \text{cm}^3 \ ^\circ\text{k mole}^{-1}}{6.090 \times 10^{34} \text{erg}^{-1} \ ^\circ\text{K mole}^{-1}} = 2.66 \times 10^{-36} \text{ esu}^2 \text{ cm}^2$$

$$\mu = [2.66 \times 10^{-36} \text{esu}^2 \text{ cm}^2]^{1/2} = 1.63 \times 10^{-18} \text{eus cm}$$

$$\times 1 \times 10^{18} \frac{\text{D}}{\text{esu cm}}$$

and therefore $\mu = 1.63$ D.

Q. 17 : Would you expect $He_2^{\oplus}$ to be more stable than He_2? than He ?

Solution : This problem requires a consideration of bonding and antibonding orbitals. An atomic orbital may be thought of as a particular volume element at a given distance and direction from the nucleus where there is a probability of finding an electron. According to the Pauli exclusion principle, no more than two electrons can occupy an orbital. Two overlapping orbitals can be considered to combine to give one low-energy bonding orbital and one high-energy antibonding orbital.

Consider the electron configuration of a helium atom (He) : $1s^2$. We can readily see that the outermost shell of helium, the s orbitals, is filled with 2 electrons, which is the maximal and stable capacity. Thus the helium atom is very stable. In fact, it is inert to most chemical reactions and is for this reason known as one of the inert gases.

To form He_2, each helim atom contributes 2 ls electrons for a total of 4 electrons. Since the molecular bonding orbital can accommodate only 2 electrons, the remaining 2 must go into the antibonding orbital. Hence, any stabilization gained by filling the bonding orbital is offset by the required filling of the antibonding orbital. Hence, there exists no net decrease in energy when He_2 is produced, so that He_2 is less stable than He.

The only difference between He_2 and $He_2^{\oplus}$ is that the latter has only 1 electron (not 2) in the antibonding orbital. It possesses two electrons in the bonding orbital as does He_2, but because of the one less electron in destabilizing antibonding orbital in $He_2^{\oplus}$, it should be more stable than He_2. (It is, of course, less stable than He which has no electrons in the antibonding orbital).

Q. 18 : Consider an octahedral complex having a univalent negative ion at each vertex of the octagon. Explain why the $d_{x^2-y^2}$ orbital of the central atom is less stable relative to the d_{xy} orbital.

Solution : Both the $d_{x^2-y^2}$ and the d_{xy} orbitals are centered about the x–y plane; hence, the difference in the stability of the two orbitals must be related to thier different orientations along the x– and y–axes.

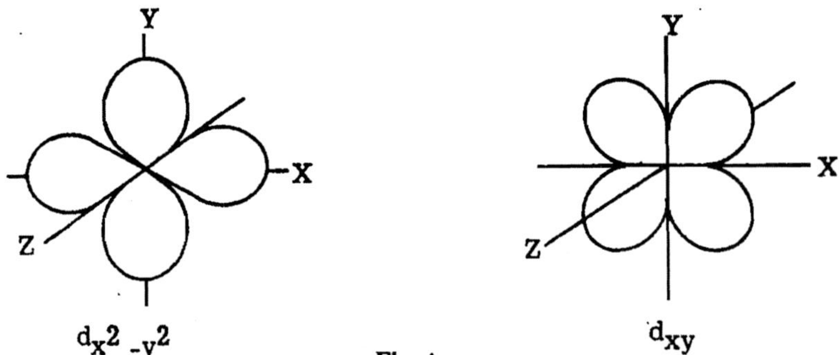

Fig. A

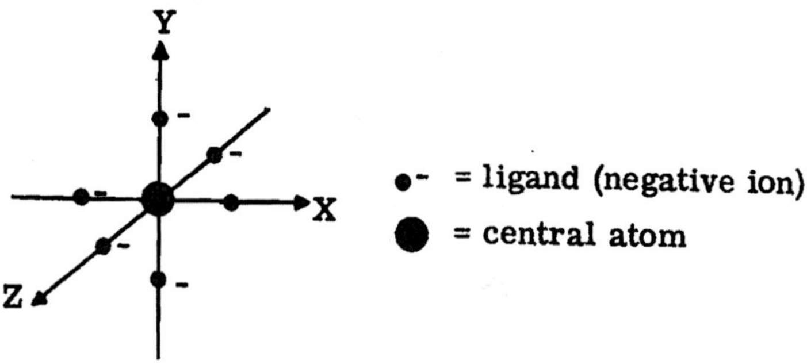

Fig. B

In the $d_{x^2-y^2}$ orbital, the lobes are directed along the axes; whereas in the d_{xy} orbital, the lobes are directed along the bisectors of the angles between the axes (see Fig. A). In the octahedral configuration the ligands (the negatively charged ions electrostatically bound to the central atom) lie on the axes (see Fig. B).

Hence, the $d_{x^2-y^2}$ orbital brings the negatively charged electrons in the lobes closer to the negatively charged ligands than does the d_{xy} orbital. Since the negative charge of the electrons repels that of the ligands, and this electrostatic repulsion falls off as the reciprocal of the square of the distance between the charges, the greater degree of repulsion between the $d_{x^2-y^2}$ orbital and the ligands, relative to the repulsion between the d_{xy} orbital and the ligands, decreases the stability of the $d_{x^2-y^2}$ orbital.

Q. 19 : How would one expect the bond strength of NO to compare with that of O_2?

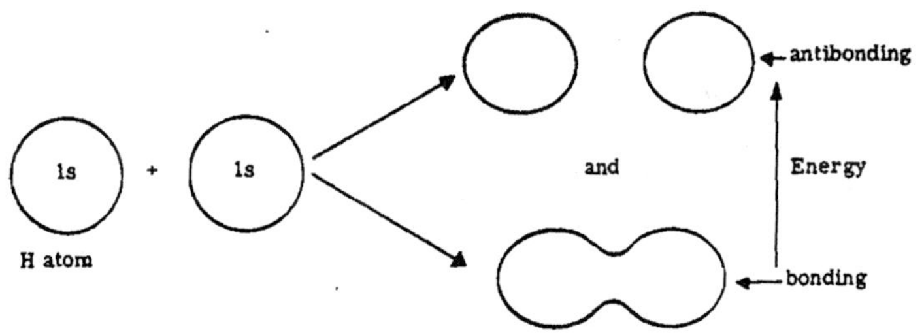

Solution : The molecular orbital is the sum of the atomic orbitals. For example, when two atomic orbitals are combined, two molecular orbitals are formed. One orbtial is called a bonding orbital and the other is an antibonding orbital. The bonding orbital is at a lower energy level than the antibonding orbitla. If possible, electrons seek out the bonding orbtial rather than the antibonding orbital. The orbitals in the production of H_2 from H atoms can be visualized as shown in accompanying figure.

The greater the number of antibonding orbitals in a molecule, the weaker the bond. Bond order is a quantity that indicates the strength of bonding orbitals. It is defined as half the number of bonding electrons minus half the number of antibonding electrons. Thus, the higher the bond order, the stronger the chemical bond.

To comapre the bond strength of NO with O_2, compare their bond orders. To do this, consider the total number of valence electrons in each element. Valence electrons are the outer electrons, which participate in bonding. For NO, the total number of valence electrons is $3 + 4 = 7$. There exist 3 bonding p orbitals, which accommodate 6 electrons. The 1 unpaired electron must be in an antibonding orbital. This means that the bond order of NO is ½ (6) - ½ (1) = 3-0.5 = 2.5.

In O_2 there is a total of 8 valence electrons. This means 2 electrons must be in antibonding orbitals, since the 3 bonding orbitals can accommodate only 6 electrons. The bond order of O_2 = ½ (6) – ½ (2) = 2. The bond order of NO is higher, which means, its chemical bond is stronger than that of O_2.

6

Periodic Properties

Q. 1 : What is the meaning and significance of the Pauling electronegativity scale ?

Solution: In any chemical bond, electrons are shared between the bonding atoms. In covalent bonds, the valence electrons are shared almost equally-the shared electrons spend about the same amount of time with each atom. In ionic bonds, the valence electrons are monopolized by one atom. The degree to which a bond will be ionic or covalent is dependent upon the relative electron-attracting ability of the bonding atoms.

The Pauling electronegativity scale provides a measure for the relative electron-attracting abilities or electronegativity of each element. The most electronegative element, fluorine, is assigned the highest number, 4.0. The least electron attracting elements (and consequently, those most willing to lose electrons), cesium and francium, are assigned the lowest number, 0.7. Numbers are assigned to the remaining elements so that no element has a higher number than an element more electronegative than it. Furthermore, the numbers are assigned so that the differences between the electronegativities of two elements is indicative of the ionic quality of the bond that forms between them. In particular, if the difference is 1.7 or greater, then the shared electrons are monopolized by the more electronegative atom to such a great extent that the bond is said to be ionic. If the difference is less than1.7, the bond is covalent.

Thus, when given a compound, the Pauling electronegativity scale can be used to determine the species that acts as the electron donor, the species that acts as the electron acceptor, and the degree of electron polarization of the bond.

Q. 2 : Classify the bonds in the following as ionic, ionic-covalent, covalent-ionic and covalent: I_2, LiCl, MgTe, Cl_2O, H_2S.

Solution: The bonds in these molecules can be classified by considering the electronegativities of the atoms involved.

The electronegativity of any element is the tendency of that element to attract electrons. The greater the electronegativity, the stronger the attraction. When 2 atoms join, the type of bond formed can be determined by calculating the difference in electronegativity of the two atoms involved. When the difference is greater than 1.7, the bond is classified as ionic, if less than 1.7, then the bond is covalent. An ionic bond involves the complete transfer of electrons from I atom to another, while in a covalent bond, then electrons are shared between like or similar atoms. Covalent–ionic indicates a mixed character which is more covalent in nature. Ionic-covalent bonds have more of an ionic character.

Consider the cases in point. For I_2, there is no difference in electronegativity because the 2 atoms involved in the bond are the same, and hence contains a covalent bond with equal sharing of the electrons.

In LiCl, the electronagetivities are Li = 1.0 and Cl = 3.0. (The electronegativity values are obtained from an electronegativity table). The electronegative difference is 2.0. Therefore, the bond is ionic.

One proceeds in a similar manner for the remainder of the compounds.

Compounds and Electronegativities of the elements	Difference in Electronegativity	Type of Bond
Mg – Te 1.2 2.1	2.1 – 1.2 = 0.9	Covalent–ionic
Cl_2 – O 3.0 3.5	3.5 – 3.0 = 0.5	Covalent–ionic
H_2 – S 2.1 2.5	2.5 – 2.1 = 0.4	Covalent–ionic

Q. 3 : In a given chemical compound, there is a bond between "A" and "B". The electronegativities of the involved elements are, respectively, 1.2 and 2.1 what is the probable nature of the bond between "A" and "B" ?

Solution: Electronegativity is a measure of the tendency of an atom to attract shared electrons in a molecule. The greater the difference in electronegativity between two elements, the greater the ionic character of the bond. When the difference in electronegativities is greater than 2.0, the bond is considered ionic.

The electronegativity difference in this problem is 2.1 – 1.2 or 0.9. Thus, while the bond is not ionic, the difference suggests a degree of polarity in this covalent bond.

Q. 4 : What is the difference between electronegativity and electron affinity ? Use specific examples.

Solution: Electron affinity represents the energy released when an electron is added to an isolated neutral gaseous atom. Electronegativity is the tendency of an element to attract electrons, not in the gaseous state, but from a nucleus joined to it by a covalent bond. The electron affinity is a precise quantitative term while the electronegativity is a more qualitative concept.

Q. 5 : Assuming the ionization potential of sodium is 5.1 eV and the electron affinity of chlorine is 3.6 eV, calculate the amount of energy required to transfer one electron from an isolated sodium (Na) atom to an isolated chlorine (Cl) atom.

Solution: Ionization potential is the amount of energy required to pull an electron off an isolated atom. Electron affinity is the amount of energy released when an electron is added to an isolated neutral atom. In this problem, one must add energy to remove an electron from Na and energy will be released upon the addition of an electron to Cl. 5.1 eV are needed to expel an electron from Na and 3.5 eV are released when Cl accepts an electron. Thus, the amount of energy required for the overall process to occur is the difference between the ionization potential of Na and the electron affinity of Cl.

The energy necessary to be added to the system for this reaction to occur is 5.1 eV – 3.6 eV or 1.5 eV.

Q. 6 : It is a fact that the second ionization potential of alkali atoms falls off more rapidly with increasing atomic number than does the first ionization potential. Why ?

Solution: The first ionization potential is the energy necessary to remove one electron from a neutral atom in the gaseous state. After this first electron is removed, the atom become an ion with a +1 ionization state. Second ionization potential refers to the energy necessary to remove an electron from the +1 ion in the gaseous state.

As the atomic numbers of elements increase the atomic radii of the atoms increase. The atomic radius determines the distance from the nucleus to the outermost electron of the atom.

The first and second ionization potentials should fall off as the atomic radius increases. As the radius increases, the outer electron, which is the one that will be removed, is at a greater distance from the nucleus. Thus, the "pull" on it by the nucleus, is less than that of an electron closer to the nucleus. Hence, less energy should be required to remove an electron that is far away from the nucleus than one that is close.

In considering the second ionization potential, one has an added factor involved; the element is charged. The concentration of charge will be decreased as the atomic radius increases. This increased diffusion of charge with the increased atomic number is the reason for the added fall off of the second ionization potential as compared to first ionization potential.

Q. 7 : 40.0 kJ of energy is added to 1.00 gram of magnesium atoms in the vapor state. What is the composition of the final mixture? The first ionization potential of Mg is 735 kJ/mole and the second ionization potential is 1447 kJ/mole.

Solution: Ionization potential may be defined as the energy required to pull an electron away from an isolated atom. The second ionization potential is the amount of energy required to pull off a second electron after the first has been removed.

The composition of the final mixture is determined by calculating the number of electrons that will be removed from the magnesium ions. To do this one must determine the number of moles of Mg present in 1 g. From this one can determine the number of electrons that will be liberated by using the values for the first and second ionization potentials of Mg.

The atomic weight of Mg is 24.3. Since moles = grams/atomic weight, there are in 1 gram of Mg, 1/24.30 or 4.11×10^{-2} moles present. The first ionization potential of Mg is 735 kJ/moles. Therefore, 4.11×10^{-2} moles of Mg requires 4.11×10^{-2} moles $\times$ 735 kJ/mole or 30.2 kJ to ionize all of the atoms once. 40 kJ was added to the system leaving 40 kJ – 30.2 kJ or 9.8 kJ to remove the second electron. If one has 9.8 kJ and 1447 kJ/mole is required to remove the second electrons, then

$$\frac{9.8 \text{ kJ}}{1447 \text{ kJ / mole}} = 6.77 \times 10^{-3} \text{ moles of atoms can}$$

have their second electron removed. 4.11×10^{-2} moles of Mg are present.

$$\frac{6.77 \times 10^{-3}}{4.11 \times 10^{-2}} \times 100 = 16.5\%.$$

This means that 16.5% of the atoms can have a second electron removed. Therefore, the composition of the mixture is: Mg^{++} 16.5%, mg^{+} 100 – 16.5 or 83.5%.

Q. 8 : Consider the formation of an ionic molecule, AB, by the donation of an electron from atom A to atom B. If the ionization potential of A is 5.5 eV and the electron affinity of B is 1.5 eV, and the ion B^{-} has a radius of 0.20 nm, what is the maximum size of A^{+} that would lend itself to the formation of an energetically stable bond?

Solution: When two atoms form an ionic bond, the energy difference between the ionization potential of one and the electron affinity of the other goes into the electrostatic attraction between the ions of the two atoms. This energy difference is related to the internuclear separation between ionized atoms by the following relationship :

$$E = 1.44 \times \frac{q_1 q_2}{r}$$

where E is the energy difference in electron-volts (eV), q_1 and q_2 are the charges on the ions in electrostatic units (esu), and r is the internuclear separation in nanometers (nm). In this problem we can determine E and use this to find r.

When A ionizes to A^+, it absorbs 5.5 eV of energy. When B accepts an electron to form B^-, it gives up 1.5 eV of energy. Hence, a total energy of $E = 1.5 \text{ eV} - 5.5 \text{ eV} = -4.0 \text{ eV}$ goes into the formation of the AB ionic bond.

The charge on A^+ is $q_1 = +1$ esu and that on B^- is $q_2 = -1$ esu. One can now determine the internuclear separation of this bond by solving the equation

$$E = 1.44 \times \frac{q_1 q_2}{r} \quad \text{for r to obtain :}$$

$$r = \frac{1.44\ q_1\ q_2}{E} = \frac{(1.44)(1 \text{ esu})(-1 \text{ esu})}{-4.0 \text{ eV}} = 0.36 \text{ nm.}$$

Hence, the nucleus of A is separated from the nucleus of B by 0.36 nm. Since the radius of B^- is 0.20 nm, the maximum radius of A^+ (at the point where it just "touches" B^-) is 0.36 nm − 0.20 nm = 0.16 nm.

Q. 9 : A chemist has one mole of X atoms. He finds that when half of the X atoms transfer one electron to the other half, 409 kJ must be added. If all of the resulting X^- ions are subsequently converted to X^+ ions, an additional 733 kJ must be added. Find the ionization potential, and the electron affinity of X.

Solution: Two facts must be known before this problem can even be attempted :

(1) The ionization potential is the energy required to pull off an electron from an isolated atom, and (2) electron affinity is the energy released when an electron adds to an isolated neutral atom. In this problem, there are two different reactions. The first is one half of a mole of neutral atoms losing electrons, and one half of a mole of neutral atoms gaining electrons. The amount of energy needed for this process to occur is the amount of energy expended to remove the electrons minus the amount of energy released by the atoms accepting the electrons. Energy required = ionization potential of X atoms – electron affinity of X atoms. Let x = ionization potential of entire mole, and let y = electron affinity of entire mole. It is given that the energy required for this process is 409 kJ and that only half the mole of atoms is involved. Thus, one can write

$$\frac{1}{2}x + \frac{1}{2}y = 409 \text{ kJ.}$$

In the second process, one half of a mole of X^- ions becomes X^+ ions. To do this, one must remove an electron from each X^- ion to form an X atom. This amount of energy needed for this process to occur is equal the opposite of the electron affinity. An electron from the X atom is then removed to form the x^+ ion. The amount of energy necessary for this reaction to occur is equal to the ionization potential. The sum of the reverse of the electron affinity and the ionization potential is equal to the energy for the overall process required. One can therefore write :

$$\frac{1}{2}x+\frac{1}{2}y=733$$

The two processes in this problem are represented by two equations using two variables. These equations are solved simultaneously by adding them together.

$$\frac{1}{2}x-\frac{1}{2}y=409$$

$$\frac{1}{2}x+\frac{1}{2}y=733$$

$$x = 1142 \text{ kJ/mole}$$

Substituting this value back into one of the equations, one can solve for y. y = 324 kJ/mole. 96.48 kJ/mole = 1 eV/particle. Therefore,

$x = 11.84$ eV/atom, $y = 3.36$ eV/atom.

Q. 10 : What is the explanation for the following trends in lattice energies ?

NaF	– 260 Kcal/mole	NaCl	– 186 Kcal/mole
NaCl	– 186 Kcal/mole	KCl	– 169 Kcal/mole
NaBr	– 177 Kcal/mole	CsCl	– 156 Kcal/mole

Solution: Lattice energy is an indication of the stability of the species.

In the first column, since the sodium (Na) ion is common to all species, one should concentrate of F^-, Cl^-, and Br^- ions. It there any change in any property when these ions are compared ? Notice that in going from F^- to Cl^- to Br^-, there is an increase in ionic size. The same situation exists in the second column in going from Na+ to K^+

to Cs^+, with Cl^- being common to all. How does this relate to lattice energy ? The energy derived from two ions in a compound stems from the mutual attraction of the oppositely charged ions. The closer the ions, the greater the attraction and the larger the lattice energy. In each column in this problem, as the size of one ion increases, the ions in the lattices become further apart. As such, there is a decrease in attraction between adjacent ions and, thus, a lowering of the magnitude of the lattice energy.

Q. 11 : Write the Born-Haber cycle for the formation of crystalline sodium fluoride (Na^+F^-), starting with solid Na and gaseous F. Then, using the thermochemical data supplied below, determine its heat of formation :

(1) Na (s) $\rightarrow$ Na (g) $\Delta H = +26.0$ Kcal : sublimation

(2) F_2 (g) $\rightarrow$ 2F (g) $\Delta H = +36.6$ Kcal : dissociation

(3) Na (g) $\rightarrow Na^+$ (g) + e^- $\Delta H = +120.0$ Kcal : ionization

(4) F (g) + e^- $\rightarrow F^-$ (g) $\Delta H = -83.5$ Kcal : electron addition

(5) Na^+ (g) + F^- (g) $\rightarrow Na^+, F^-$ (s) $\Delta H = -216.7$ Kcal: lattice formation.

Solution: The Born-Haber cycle shows that the heat of the reaction is the sum of the ΔH's for the sequence of the reactions used in the formation of crystalline sodium fluoride.

The Born–Haber cycle relates reactions which produce the gaseous atoms from solids, those that produce the gaseous ions, and those that indicate the union of these ions to form the crystalline product. Thus, one can write the Born-Haber cycle for the reaction as

$$\begin{array}{ccc} Na(s) + \frac{1}{2}F_2(g) & \xrightarrow{(1)+(2)} & Na(g) \quad + \quad F(g) \\ \downarrow & & (3)\downarrow \qquad (4)\downarrow \\ Na^+,F^-(s) & \xleftarrow{(5)} & Na^+(g) \quad + F^-(g) \end{array}$$

(The numbers above the arrows correspond to the numbered reactions in the problem.)

One is given the energies, i.e, the thermodynamic data, for these processes. If one numbers the reactions in this problem (1) to (5), the reaction becomes (1) + ½ (2) + (3) + (4) + (5). The ½ value for (2) is derived from the fact that one mole of NaF is required, but, in (2), 2 moles of F(g) atoms are generated. Thus, take ½ this number.

Therefore, heat of reaction is :

$\Delta H = \Delta H\ (1) + \frac{1}{2}\Delta H\ (2) + \Delta H\ (3) + \Delta H\ (4) + \Delta H\ (5)$

$= 26 + 18.3 + 120 - 83.5 - 216.7 = -\ 135.9$ Kcal/mole.

Q. 12 : Calculate the interatomic nonbonded distance between 2 bromine nuclei in a molecule of carbontetrabromide (CBr_4). The C–Br distance is 1.94 Å.

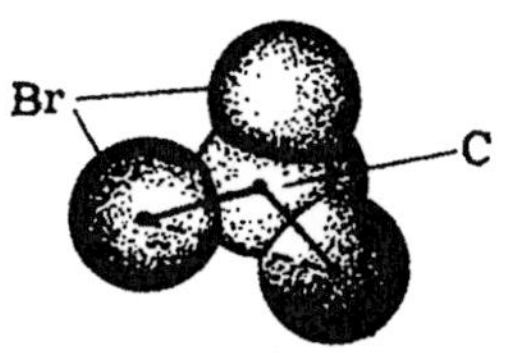

Diagram A

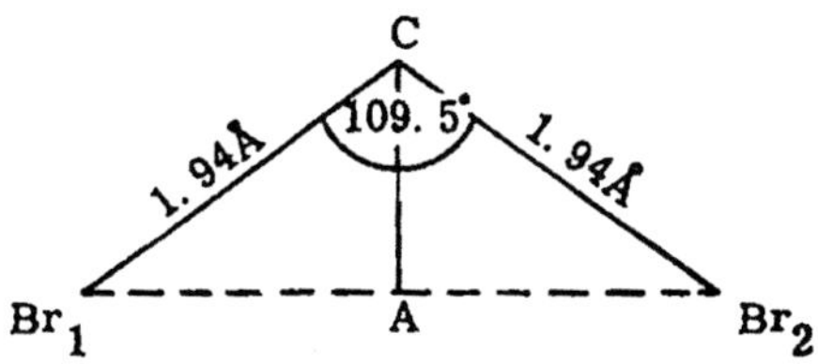

Diagram B

Solution: From diagram A, one can see that CBr_4 is in a tetrahedral conformation. Thus, one knows that the Br–C–Br angle is equal to 109.5°. As such, diagram B is then true.

One is solving for the length of B $\overline{Br_1 Br_2}$.

Angle $Br_1 - C - A$ is 109.5°/2 or 54.75°, since one can let line $\overline{CA}$ be an angle bisector. To find $\overline{Br_1A}$, note that sin 54.75 = B_1A/ 1.94 Å. Therefore,

$$\overline{Br_1A} = 1.94\ \text{Å} \times \sin 54.75$$

$$\overline{Br_1A} = 1.94\ \text{Å} \times .8166 = 1.58\ \text{Å}$$

$$\overline{Br_1B_2} = 2\ \overline{Br_1A} = 2 \times 1.58\ \text{Å} = 3.17\ \text{Å},$$

since $\overline{Br_1A} = \overline{Br_2A}$ for $\overline{CA}$ is also a median.

Q. 13 : Given 0.207 nm for the S–S distance and 105° for the S–S–S bond angle, calculate the S–to–S distance across the S_8 ring.

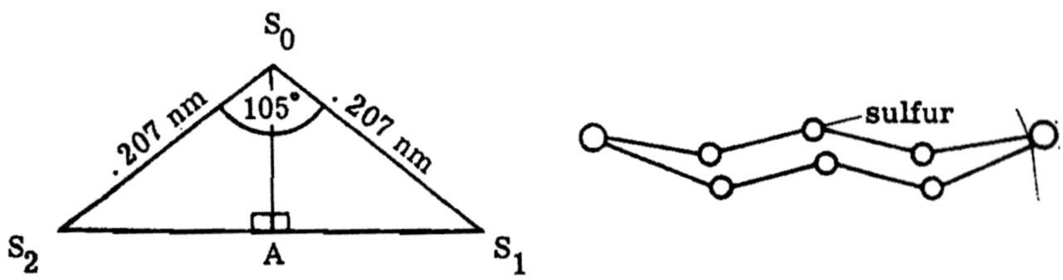

Fig. A-B

Solution: One is aked to find the distance from S_1 to S_2 as seen in figure A. This can be done by using trigonometry. Draw a line perpendicular to S_1S_2 to A from S_0. One can solve for the length of S_2A with the knowledge that the angle S_2-S_0–A is equal to ½ of the S–S–S angle or 52.5°. This is true because the triangle $S_0S_1S_2$ is isosceles.

$$\sin 52.5° = \frac{S_2A}{S_2S_0} = 0.79$$

$$S_2A = 0.793 \times S_2S_0 = (0.793) \times (0.207\ \text{nm}) = 0.164\ \text{nm}.$$

$$S_2S_1 = 2 \times S_2A = 2 \times 0.164\ \text{nm} = 0.328\ \text{nm}.$$

Q. 14 : Given that the H–to–H distance in NH_3 is 0.1624 nm and N–H distance is 0.101 nm, calculate the bond angle H–N–H.

Solution: Because the NH distances are equal, an isosceles triangle can be drawn as shown. The altitude NX is drawn to form two right triangles. The magnitude of the angle HNX can be found trigonometrically from its sine. The sine of an angle is equal to the length of the opposite side divided by the hypotenuse.

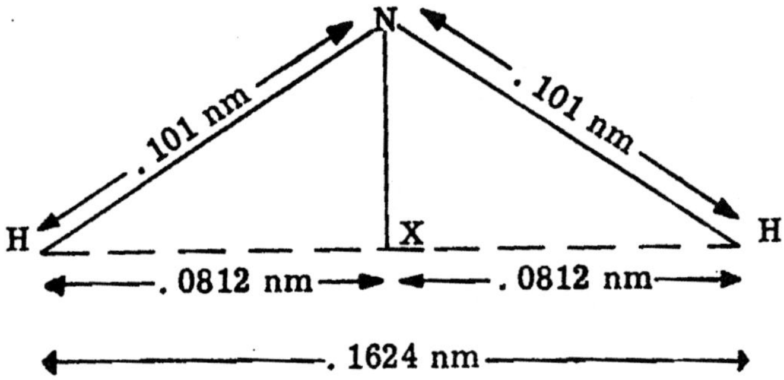

Fig. A

$$\sin HNX = \frac{0.0812 \text{ nm}}{0.101 \text{ nm}} = 0.804.$$

Using a table of trigonometric functions, one can find the angle whose sin is 0.804.

$\text{Sine}^{-1}\ 0.804 = 53.50°$.

The angle H–N–H is twice HNX. As such, the measure of angle H–N–H is 2 × 53.50° or 107°. It is known that H–N–H is twice HNX because the altitude of an isosceles triangle bisects the vertex angle.

Q. 15 : In the trans form of nitrous acid, all the atoms are in the same plane, but the HO is directed away from the other O. The O–H distance is 0.098 nm; the distance from the central N to the hydroxyl O is 0.146 nm; and the distance to the other O is 0.120 nm. If the H–O–N bond angle is 105° and the O–N–O bond angle is 118°, how far is the H from the other O ?

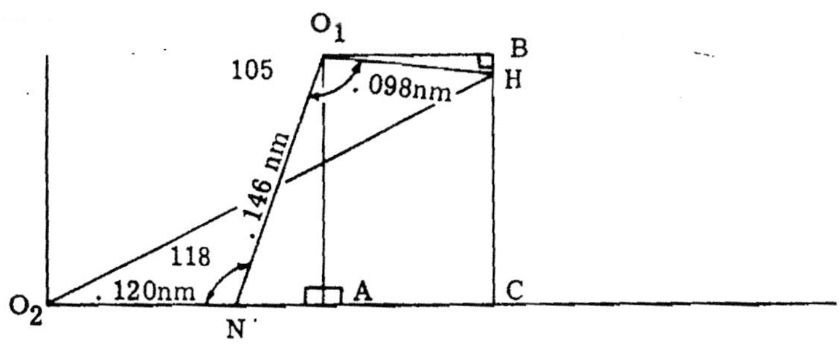

bond distance - molecular structure

Fig.

Solution: The formula for nitrous acid in HNO_2. The structure of the trans form as described in the problem is :

One is asked to find the distance from O_2 to H. This can be done by solving for the length of O_2C and CH. The sum of the squares of these two lenghts is equal to the square of the length of O_2H. This can be done through the use of trigonometry. To find O_2C, one must find NA and AC. O_2C is equal to the sum of O_2N, Na, and AC. The measure of angle ONA is equal to 180° – 118° or 62°, since they are supplementary.

$$\cos 62° = 0.469 = \frac{NA}{NO_1} = \frac{NA}{0.146\text{ nm}}\left(\frac{\text{adjacent}}{\text{hypotenuse}}\right)$$

$$0.469 = \frac{NA}{0.146\text{ nm}}$$

$$NA = 0.069\text{ nm}$$

$$O_2A = NA + O_2N = 0.069\text{ nm} + 0.120\text{ nm} = 0.189\text{ nm}$$

$\overline{AC} = \overline{O_1B}$, since they are the opposite sides of a rectangle.

$\angle BO_1H = 90° - \angle HO_1A$ ($\angle$ = angle)

$\angle NO_1A = 90° - \angle O_1NA = 90° - 62° = 28°$

$\angle HO_1A = \angle HO_1N - \angle NO_1A = 105° - 28° = 77°$

$\angle BO_1H = 90° - 77° = 13°$ from above.

$$\cos 13° = \frac{\overline{OB}}{\overline{OH}} = 0.974$$

$$\frac{\overline{O_1B}}{\overline{O_1H}} = \frac{\overline{O_1B}}{0.098\text{ nm}} = 0.974$$

$$\overline{O_1B} = (0.098\text{ nm}) \times (0.974) = 0.095\text{nm}$$

Thus, $\overline{AC} = 0.095\text{ nm}. \overline{AC} = 0.095\text{ nm}.$

Solving for $\overline{O_2C}$:

$$\overline{O_2C} = \overline{O_2N} + \overline{NA} + \overline{AC}$$

$$= 0.120\text{ nm} + 0.069\text{ nm} + 0.095\text{ nm} = 0.284\text{ nm}$$

$$\overline{HC} = \overline{OA} - \overline{BH}$$

$$\text{Sin } \angle ONA = \frac{\overline{O_1A}}{\overline{NO_1}} \qquad \angle ONA = 62°$$

$$\sin 62° = \frac{\overline{O_1A}}{\overline{NO_1}} = 0.883$$

$$\frac{\overline{O_1A}}{\overline{NO_1}} = 0.883$$

$$\overline{OA} = (0.883) \times \overline{NO_1} = (0.883) \times (0.146\text{nm})$$

$$= 0.129 \text{ nm}$$

$$\sin \angle BOH = \frac{\overline{BH}}{\overline{O_1H}} \qquad \angle BOH = 13°$$

$$\sin 130° = \frac{\overline{BH}}{0.098 \text{ nm}} = 0.225$$

$$\overline{BH} = (0.098 \text{ nm}) \times (0.225) = 0.022 \text{ nm}$$

$$\overline{HC} = 0.129 \text{ nm} - 0.022 \text{ nm} = 0.107 \text{ nm}$$

$$\left(\overline{O_2H}\right)^2 = \left(\overline{O_2C}\right)^2 + \left(\overline{HC}\right)^2$$

$$\left(\overline{O_2H}\right)^2 = (0.284 \text{ nm})^2 + (0.107 \text{ nm})^2$$

$$\overline{O_2H^2} = 0.081 \text{ nm}^2 + 0.011 \text{ nm}^2$$

$$= 0.092 \text{ nm}^2$$

$$\overline{O_2H} = 0.303 \text{ nm.}$$

Q. 16 : Find the net dipole moment in Debyes for each of the following situations : (a) One + 1 and one – 1 ion separated by 2×10^{-8} cm, and (b) one + 2 and one – 2 ion separated by 2×10^{-8} cm. One charge = 4.80×10^{-10} esu and 1 Debye = 1×10^{-18} esu–cm.

Solution: A dipole consists of a positive and a negative charge separated by a distance. Quantitatively, a dipole is described by its dipole moment, which is equal to the charge times the distance between the positive and negative centers. For both situations, substitute the values into this quantitative expression :

net dipole moment = charge × distance

(a) Net dipole moment

$$= (4.80 \times 10^{-10}\ \text{esu})\ (2 \times 10^{-8}\ \text{cm})$$

$$= 9.6 \times 10^{-18}\ \text{esu–cm.}$$

1 Debye $= 1 \times 10^{-18}$ esu–cm.

$$\frac{9.6 \times 10^{-18}\ \text{esu} - \text{cm}}{1 \times 10^{-18}\ \text{esu} - \text{cm / Debye}} = 9.6\ \text{Debyes}$$

(b) Net dipole moment

$$= (2)\ (4.80 \times 10^{-10})\ (2 \times 10^{-8})$$

$$= 19.2 \times 10^{-18}\ \text{esu–cm}$$

Converting to Debyes

$$\frac{19.2 \times 10^{-18}\ \text{esu} - \text{cm}}{1 \times 10^{-18}\ \text{esu} - \text{cm / Debye}} = 19.2\ \text{Debyes.}$$

7

Representative Elements (s–and p-block elements)

Q. 1 : Explain why are alkali metals electrically conducting ?

Ans. Good electrical conductivity of alkali metals is due to the presence of a *metallic bond* in their crystal lattice, in which cations (or kernels) are held in fixed positions in space; while the free and mobile electrons can move under the influence of applied electric field.

Q. 2 : Why are group I metals called alkali metals ?

Ans. Group I metals are called alkali metals, because their hydroxides (*e.g.*, NaOH, KOH) are strong alkalies.

Q. 3 : Alkali metals are difficult to reduce. Explain why.

Ans. Alkali metals are highly electropositive metals, because they possess strong tendency to lose the valence electron, owing to their low ionization energy values. Thus, they possess great tendency to be oxidised readily. In other words, alkali metals are *not reduced,* because addition of electron is not favoured energically.

Q. 4 : Why are alkali metals strong reducing agents ?

Ans. Alkali metals are strong reducing agents because :

(*i*) they readily lose the lone valence electron, owing to their low ionization energy values.

$$Na\,(g) \longrightarrow Na^{+}(g) + e^{-}$$

(*ii*) of their low heat of atomization; and

$$Na\,(s) \longrightarrow Na\,(g)$$

(*iii*) of high heat of hydration.

Q. 5 : Exaplain why alkali metals form soluble hydroxides, which become stronger alkalies on descending the group.

And. (*i*) **Soluble salts :** The alkali metals form large-sized monovalent cation (*e.g.*, M^+), so their compounds possess low value of interionic attractions. So their compounds have low lattice energy; and the lattice energies are readily overcome by the ion-dipole interactions with polar water molecules. Hence, their salt (like hydroxides) are readily soluble in water.

(*ii*) **Alkaline nature of hydroxides :** On descending the group, the size of alkali metal cation increases, thereby the interionic attraction forces decrease. Consequently, it becomes more easy for the alkali metal hydroxide to release its hydroxyl ion. Hence, the alkaline character of alkali metal hydroxides increases on descending the group.

Q. 6 : Why do beryllium halides fume in air ?

Ans. Be^{2+}ion has a strong tendency to form hydrates, since it has a high value of hydration energy. Consequently, beryllium halide is *highly hygroscopic* and it absorbs moisture (of air) and hydrolyses with the formation of fumes of hydrogen halide. Thus :

$$BeCl_2 + 2\,H_2O \longrightarrow Be\,(OH)_2 + 2\,HCL \uparrow$$

Q. 7 : Alkali and alkaline-earth metals cannot be obtained by chemical reduction methods. Explain why.

Ans. Alkali and alkaline-earth metals are *strong reducing agents.* So they cannot be obtained in the metallic form by chemical reduction of their oxides or other compounds. Consequently, they are, usually, obtained by electrolysis of their fused anhydrous salts.

Q. 8 : State the following about boron family (Group 13 elements):

(*i*) Tendency to hydride formation, and

(*ii*) Stability of oxidation states.

Ans. Boron familiity (group 13 elements):

(*i*) **Tendency of hydride formation** *decreases down the group.* Thus, boron forms stable covalent hydrides, called *boranes* such as B_nH_{n+4} and B_nH_{n+6}. Other members form only a few stable hydrides such as AIH3 (a colourless solid).

(ii) **Stability of oxidation states** + *3 oxidation state decreases in the sequence :*

$$Al < Ga < In < Tl,$$

On the other hand, + *1 oxidation assumes impotance as we descend the group from Al to Tl.* Hence, + 1 oxidation state for Tl is more stable than + 3 oxidation state in aqueous solution (*inert-pair effect*).

Q. 9 : Why does aluminium not react with cold water under ordinary conditions ?

Ans. Due to the formation of a very thin, transparent, non-porous, protective and tenaceously-adhering film of aluminium oxide (Al_2O_3) on its surface, aluminium does not react with cold water under ordinary conditions.

Q. 10 : "Aluminium cannot be extracted by reduction of its ores by carbon". Explain briefly.

Ans. Aluminium is a strongly electropositive metal. So its ore, alumina (Al_2O_3), cannot be reduced by carbon, during the extraction of aluminium. Moreover, carbon immediately combines with aluminium, if formed, to yield aluminium carbide (Al_4C_3).

Q. 11 : "Aluminium is more active than iron and yet there is less corrosion of aluminium, when both are exposed to air. Explain.

Ans. Due to the formation of a thin, transparent, protective, non-porous and tenaceously-adhering film of aluminium oxide (Al_3O_2) on its surface, aluminium is practically corrosion-resistant to air as well as moisture. On the other hand, iron is corroded in moist air, due to formation of porous, non-adhering layer of the rust ($Fe_2O_3xH_2O$).

Q. 12 : Boron does not form B^{3+}ion.

Ans. Boron mainly forms covalent bonds, because of its small-sized atom and high ionization energy, due to which it cannot lose electrons. Hence, boron does not form B^{3+}ion.

Q. 13 : Boron trifluoride is a strong Lewis acid. Explain.

(*Delhi Board, 1986, 1979*)

Ans. In boron trifluoride (BF_3) molecules, three electrons of boron atom share with three electrons, one each from three fluorine atoms. Thus, the total number of electrons in the outermost shell of boron atom in BF_3 is six, which is short of two electrons than nearest inert gas, neon. Thus, boron trifluoride is an *'electron-deficient'* compound. Therefore, BF_3 has a strong tendency to accept electrons from an electron-rich compound (*i.e.*, a Lewis base). Consequently, BF_3 behaves as a strong Lewis acid. Thus:

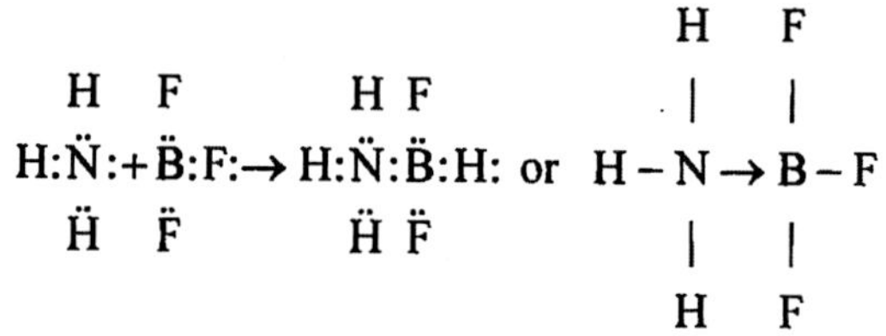

(Electro-rich)	(Election-deficient)	(Adduct)
Ammomin	Boron trifluoride	
(Base)	(Acid)	

Q. 14 : Explain the following observations :

(*i*) NH_3 has a higher boiling point than PH_3.

(*Delhi Board, 1979, 1978*)

(*ii*)Ammonia is a good complexing agent.

(*Delhi Board, 1979*)

Ans. (*i*) Ammonia (mol. mass = 17) should have lower "boiling point" than phosphine (mol. mass = 34). However, ammonia actually has a 'higher boiling point' than phosphine.

Reason : Nitrogen is *highly electronegative* and *small in size,* so molecules of ammonia get *associated through hydrogen bonding, thereby the effective molecular mass of ammonia molecules become much higher than that of phosphine,* in which there is no hydrogen bonding, since phosphorus atom is quite large and much less electronegative. *Hence, ammonia has a higher boiling point than phosphine.*

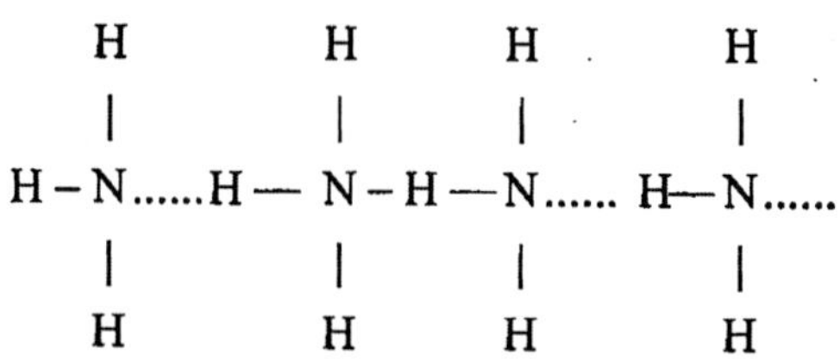

Hydrogen bond

(*ii*) Nitrogen atom (N) in ammonia (NH_3) molecule possesses a *lonepair of electrons,* which is readily available for co-ordinate bond formation with cations hving vacant orbitals. Thus, ammonia readily forms complex with silver ions and cupric ions.

$$2\,H_3N: + Ag^+ \longrightarrow [Ag\,(NH_3)_2]^+$$

$$4\,H_3N: + Cu^{2+} \longrightarrow [Cu\,(NH_3)_4]^{2+}$$

Q. 15 : Carbon cannot form complexes ; whereas other elements of group 14 can form. Account for it.

Ans. The tendency of an element of form a complex is favoured by its :

(*i*) high charge ;

(*ii*) small- sized atom, and

(*iii*) availability of empty orbitals of suitable energy in it.

From the valence-shell configuration of carbon atom ($2s^2$, $2p^2$), it is clear that it can accomodate a maximum of eight electrons in its valence shell while forming covalent compounds like CH_4, CCl_4, etc. Since the structure $2s^2p^6$ is akin to that of an inert gas, so these compounds are quite stable. Since carbon atom does not have *d*-orbitals available in these tetra-covalent compounds, so it cannot form any complex by accomodating any move electrons.

On the other hand, in case of other elements of group 14, *d*-orbitals are available in their tetra-covalent compounds and consequently, they can form complexes by accepting electrons from atoms or ions having long pair of electrons, *e.g.*, $SiCl_4$ can form complex like $[S_iF_6]^{2-}$ by accepting two lonepairs of electrons from two F^- ions.

$$SiF_4 + 2\,F^- \longrightarrow [SiF_6]^{2-}$$

Such complexes result from sp^3d^2 hybridization.

Q. 16 : Carbon tetrachloride (CCl_4) is resistant to hydrolysis, but silicon tetrachloride ($SiCl_4$) is readily hydrolysed. Explain.

Ans. Carbon tetrachloride (CCl_4) in inert, because carbon cannot exhibit a covalence greater than four, owing to the absence of vacant 2*d*-orbitals. Thus, carbon cannot expand its octet. Moreover, C — Cl bond is quite strong.

On the other hand, silicon tetrachloride ($SiCl_4$) is readily hydrolysed, because water can attack through the vacant 3*d*-orbitals of silicon atom. Thus :

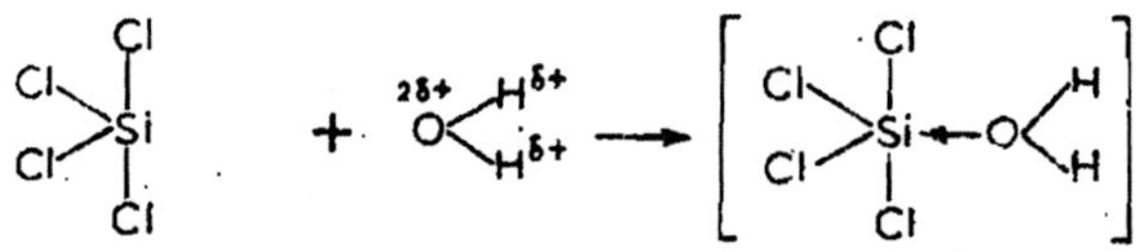

Silicon tetrachloride (Si atom has vacant 3d-orbitals)

Formation of adduct through vacant 3d-orbitals of silicon

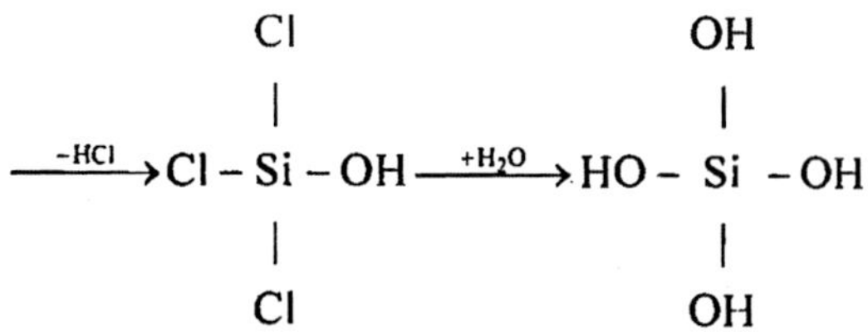

Hydrated silicic acid,
$(SiO_2\ 2H_2O)$

Q. 17 : Explain why CO_2 is a gas; whereas SiO_2 is a high melting solid.

Ans. Carbon dioxide (CO_2) is a *linear discrete molecule,* in which carbon atom is in *sp*^{3}–hybrid state.It has two *p*-bonds and is covalent in nature (O = C = O). Various molecules of CO_2 are held together by weak van der Waal's forces, which cannot cause CO_2 molecules to condense to a liquid or solid. Hence, CO_2 is a gas at ordinary temperature.

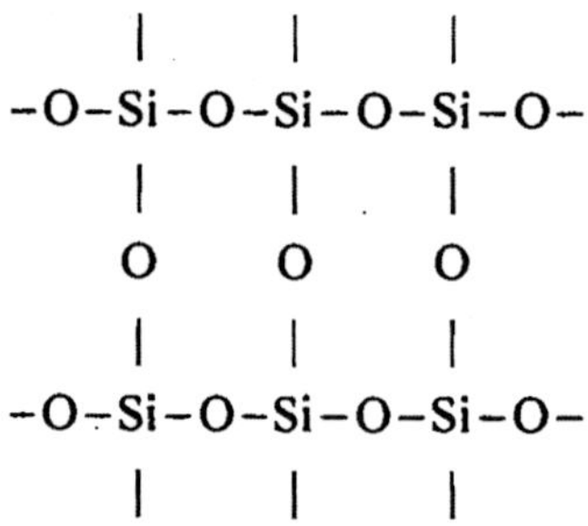

On the other hand, in silica (SiO_2), *the individual discrete molecules are not present, but it has an infinite three-dimensional structure,* in which each silicon atom is linked tetrahedrally to four oxygen atoms. Since these Si–O bonds are covalent bonds and are much stronger than van der Waal's forces, therefore, SiO_2 condenses to a solid having high melting point.

Q. 18 : Why is the property of 'catenation' shown by carbon and not to the same extent by other elements of group 14 ?

Ans. Carbon shows the property of catenation, due to its *small size* and *high electronegativity.* Since the strength of C – C bond is much more than the strength of the bonds between atoms to other elements of group 14, so elements other than carbon do not show the property of catenation to the same extent.

Q. 19 : Why is ammonia soluble in water, but other hydrides of group 15 elements are insoluble in water ?

Ans. Ammonia is soluble in water, because it forms hydrogen bonds with water. On the other hand, PH_3 and other hydrides of group 15 elements are insoluble in water, since they do not form hydrogen bonds with water molecules.

Q. 20 : Why is oxygen a gas ; while sulphur is a solid?

Ans. Oxygen forms a stable diatomic molecule, O_2. These molecules are held together by weak van der Waal's type of inter-molecular forces. Hence, oxygen exists as gas under ordinary conditions. On the other hand, sulphur forms more complex molecules, containing 8 atoms. Since the molecular mass of this complex molecule of S_8 (*see* Fig. 6) is high, so it exists as a solid.

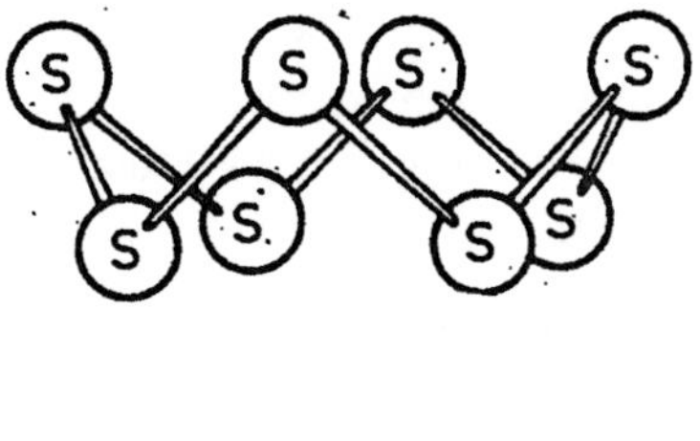

Fig.

Q. 21: Define *catenation* and illustrate it with expamples from silicon and sulphur chemistry.

Ans. Catenation *is the ability of an atom of an element to combine with itself to yield long-chain compounds, open and or closed of different sizes.*

Examples: (*i*) *Silican* shows the property of catenation and it forms Si–Si covalent bonds. The property of catenation is limited to 10 atoms of silicon.

(*ii*) *Sulphur* shows the property of catenation and it exists as S_8 molecule, in which 8 sulphur atoms are joined by single covalent bonds in the form of a pluckered octagenalring.

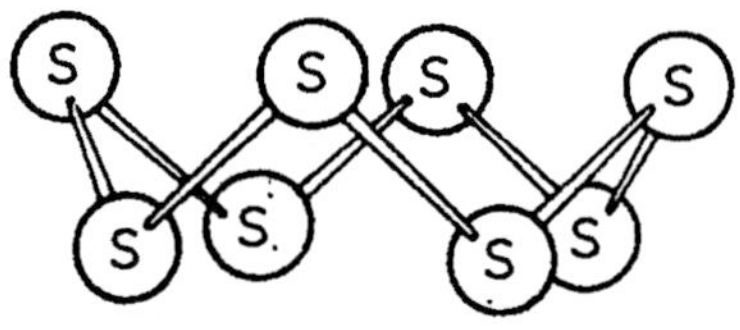

Fig.

Q. 22 : Why is OF_6 compound not known ?

(All India Board, Central Schools Delhi, 1990)

Ans. Structure of oxygen is $1s^2$; $2s^2$, $2p^4$. Since it does not have any 2*d*-orbitals in its valency shell, so it cannot expand its octet due to non-availability of d-orbitals. Hence, oxygen atom cannot exhibit higher positive oxidation state, and hence, OF_6 compound is not known.

Q. 23 : H_2O water molecule contains eight ninth of oxygen by mass, yet it is not a good oxidizing agent. Explain why.

Ans. The O — H bond in H_2O is highly stable and its breaking requires exceptionally high energy. In other words, oxygen in water is not available as free oxygen. Consequently, oxygen in water is not available for chemical reactions. Hence, water cannot act as an oxidising agent.

Q. 24: Explain, giving reason :

(*i*) Fluorine is a stronger oxidising agent than chlorine.

(*ii*) HF is a liquid ; whereas hydrides of other halogens are gases at room temperature.

(*iii*) HF liquefies more easily than HCl.

Ans. (*i*) Fluorine is a stronger oxidising agent than chlorine ? The size of fluorine atom is smaller than that of chlorine. Also the electronegativity of fluorine is much higher than that of chlorine. Consequently, fluorine is a stronger oxidising agent than chlorine. Moreover, the dissociation energy of fluorine is much lesser than of chlorine. Hence, fluorine is a stronger oxidising agent than chlorine.

(*ii*) HF is a liquid; whereas hydrides of other halogens are gases at room temperature ? Due to highest electronegativity of fluorine, HF molecules are *associated* through *hydrogen bonding.* Thus :

H—F......H—F.......H—F.....H—F.......

Consequently, the boiling point of HF is abnormally high as compared to other halogen acids. Hence, at room temperature HF is a liquid ; whereas other halogen hydrides are gases.

(*iii*) HF liquefies more easily than HCl, due to stronger hydrogen bonding in HF molecules.

......H—F......H—F......H—F......

Since fluorine is the most electronegative element, so hydrogen bonding between HF molecules is quite strong. So it liquefies easily. On the other hand, chlorine is much less electronegative than fluorine, so hydrogen bonding between HCl molecules is comparatively weaker. Consequently, HCl is not liquefied easily.

Q. 25 : Explain why fluorine does not form polyhalides ?

Ans. Electronic configuration of fluorine is $1s^2$; $2s^2$, $2p^5$. So, due to the non-availability of *d*-orbitals, fluorine cannot show higher positive oxidation states and hence, it does not form polyhalides.

Q. 26 : Why fluorine always exhibits an oxidation states of –1?

(Delhi Board, 1984)

Ans. The electronic configuration of fluorine is $1s^2$; $2s^2$, $2p_x^2$, $2p_y^2$, $2p_z^1$. Thus, it contains *one unpaired electron.* In order to complete its octet, it accepts one electron per atom to from the corresponding fluorine ion. Thus :

$$F + e \longrightarrow F^-$$

(−1 oxidation state)

Fluorine exhibits only −1 oxidation state, because *it is the most electronegative element and it cannot expand its octet, due to the non-availability of vacant d-orbitals in 2nd orbit.* The energy required to promote an electron from its 2s or 2p to 3s-orbital is very high and hence, such a possibility is also ruled out. *Hence, fluorine never exhibits a positive oxidation state.*

8

Transition Metals Including Lanthanides

Q. 1 : Which of the d-block elements are not regarded as transition elements ?

Ans. *The elemenst having completely filled penultimate d-orbital* (*i.e.*, having the configuration (n – 1) d^{10}, ns^2) *cannot be regarded as true transition elements,* since transition elements are those whose atoms or at least one of their ions have incomplete *d*-orbitals. Consequently, the elements *zinc, cadmium,* and *mercury,* belonging to the *d*-block, are not strictly regarded as transition elements. Never-the less, they are studied alongwith transition elements because they are more similar to transition elements in many respects.

Q. 2 : Zn, Cd and Hg are normally not considered transition metals. Give reason.

Ans. Outer electronic configuration of :

Zinc (Zn) : At. No. = 30 is : $3d^{10}$, $4s^2$

Cadmium (Cd) : At. No. = 48 is : $4d^{10}$, $5s^2$

Mercury (Hg) : At. No. = 80 is : 5 d^{10}, $6s^2$

Since these three elements have completely filled penultimate d-orbitals, *i.e.*, have (n – 1) d^{10}, ns^2 configuration, *so they cannot be a regarded as true transition elements, because transition elements are those whose atoms (or at least their one ion) have incomplete d-orbitals.* Consequently, d-block elements zinc, cadminm and mercury are normally not considered as transition elements.

Q. 3 : How do you explain the anomalous electronic configurations of Cr ($3d^5$, $4s^1$) and Cu ($3d^{10}$, $4s^1$) ?

Ans. Anomalous structures of Cr and Cu : A half-filled and completely full-filled orbital possesses extra-stability. So one 4*s*-electron shifts to 3*d*-orbitals, in each Cr and Cu atom, to yield extra-stable configuration of half-filled ($3d^5$, $4s^1$) and fully-filled ($3d^{10}$, $4s^1$) d-orbitals in chromium and copper respectively. Due to these extra-stable configuration, chromium and copper possess $3d^5$, $4s^1$ and $3d^{10}$, $4s^1$ configuration ; instead of $3d^4$, $4s^2$ and $3d^9$, $4s^2$ configuration respectively.

Q. 4 : In what way the electronic configurations of transition elements are different from those of representative elements ?

Ans. *Elements of representative groups (viz..* groups 1, 2, 13, 14, 15, 16, 17 and 18) *have only their outermost shell incomplete, i.e.,* they correspond to the filling up on *ns* and *np*-electrons of sub-shells in their outermost shell. On the other hand, elements of transition series (*viz.,* groups 3, 4, 5, 6, 7, 8, 9, 10, 11 and 12) *have their outermost as well as the penultimate shells incomplete.*

Gererally, they contain 2 electrons in their *ns*-orbital; while their penultimate shell (i.e., (n –1) shell, is being expended form 8 to 18 electrons by the addition of electrons in (n – 1) *d*-orbitals. Consequently, the number of electrons in their valence shell (*ns*) remains invariable the same (*i.e., ns^2*). Their general electronic configuration is: $(n-1)\ d^{1-10}$, ns^{1-2}.

Q. 5 : What are the oxidation states exhibited by lanthanide elements ?

Ans. Oxidation states of lanthanides : +3 state is common to all lanthanides. Some lanthanides also exhibit + 2 and + 4 oxidation states. However, both these states are less stable than + 3 state, because the configuration obtained is of xenon, which is highly stable. Important oxidation states of lanthanides are given below:

Element	*Symbol*	*Atomic number*	*Electronic Configuration*	*Oxidation States*
Lanthanum	La	57	[Xe]$4f^1$, $6s^2$	+3
Cerium	Ce	58	[Xe]$4f^2$, $6s^2$	+3, + 4
Praseodymium	Pr	59	[Xe]$4f^3$, $6s^2$	+3, + 4

Neodymium	Nd	60	$[Xe]4f^4, 6s^2$	+2, + 3
Promethium	Pm	61	$[Xe]4f^5, 6s^2$	+3
Samarium	Sm	62	$[Xe]4f^6, 6s^2$	+2, + 3
Europium	Eu	63	$[Xe]4f^7, 6s^2$	+2, + 4
Gadolinium	Gd	64	$[Xe]4f^7, 5d^1, 6s^2$	+3 *f*
Terbium	Tb	65	$[Xe]4f^7, 6s^2$	+3, + 4
Dysprosium	Dy	66	$[Xe]4f^{10}, 6s^2$	+3, + 4
Holmium	Ho	67	$[Xe]4f^{11}, 6s^2$	+3
Erbium	Er	68	$[Xe]4f^{12}, 6s^2$	+3
Thulium	Tm	69	$[Xe]4f^{13}, 6s^2$	+2, + 3
Ytterbium	Yb	70	$[Xe]4f^{14}, 6s^2$	+2, + 3
Lutetium	Lu	71	$[Xe]4f^{14}, 6d^1, 6s^2$	+3

Q. 6 : Why do *d*-group transition elements exhibit more oxidation states than *f*-group elements?

Ans. Both the outermost *s*-and *d*-electrons can be involved in bonding with *d*-group elements, so *d*-group transition elements exhibit more oxidation states. On the other hand, the *f*-electrons are too deeply burried in the atom to be directly involved in bonding in the lanthanides and actinides. Hence, *f*-group elements form less oxidation states.

Q. 7 : Why does not zinc show oxidation state higher than two ?

Ans. The electronic configuration of zinc is $[Ar]\ 3d^{10}, 4s^2$. So, by losing two 4*s*-electrons, it exhibits +2 oxidation state. Since the 3*d*-orbitals of zinc are ***completely filled*** and are very stable, so the 3*d*-electrons are not available for chemical formation. Hence, zinc does not show an oxidation state higher than two.

Q. 8 : Which of the two V (IV) and V (V) is paramagnetic ? Give reason for your answer. (At. No. of V = 23).

Ans. Electronic configuration of :

Vanadium (At. No. = 23) is : $[Ar]\ 3d^3, 4s^2$.

∴ Electronic configuratin of :

V^{4+}ion is : $[Ar]\ 3d^1$

and V^{5+}ion is : [Ar].

Since V^{4+} [or V (IV)] ion contains one upaired electron in 3*d*-orbitals, so it is *paramagnetic*. On the other hand, V^{5+} [or V (V)]ion is diamagnetic, since it does not contain any unpaired electron. *Hence, V (IV) is paramagnetic.*

Q. 9 : Why is Fe^{3+} more paramagnetic than Fe^{2+}?

Ans. The electronic configuration of :

Fe^{2+} = [Ar] $3d^6$;

and Fe^{3+} = [Ar} $3d^5$;

Thus, Fe^{3+} and Fe^{2+} have respectively *four* and *five* unpaired electrons in their *3d*-orbitals. Since greater the number of unpaired electrons in an atom or ion, greater will be the paramagnetism, so Fe^{3+} is more paramagnetic than Fe^{2+}.

Q. 10 : Which of the following is more paramagnetic and why ?

Cr^o and Mn^o.

Ans. The electronic configuration of :

Cr^o = [Ar] $3d^5$, $4s^1$;

and Mn^o = {Ar] $3d^5$, $4s^2$.

Since Cr has *six* unpaired electrons (5 in *3d* and 1 in 4s-orbitals), and Mn has *five* unpaired electrons (all in *3d* orbitals,), so *Cr is more paramagnetic than Mn.*

Q. 11 : Compounds of transitional elements are often coloured. Give reason.

Ans. Formation of coloured ions by transition metals : *Simple and complex compound of transition metals are, usually, coloured, both in solid form and in solution. They, generally, form coloured positive ions in solution.*

Explanation : *The colour of transition metal ions is associated with incompletely filled (n – 1) d-orbitals. The transition metal ions, containing unpaired d-electrons, undergo electronic transition from one d-orbital to another. During this transition process, they absorb certain radiations (from the visible light) and emit the remainder as coloured light. So the colour of an ion is complimentary of the colour absorbed by it. Hence, coloured ions are formed due to d-d transitions, which fall within the visible range for all transition elements.* For example, $[Cu(H_2O)_6]^{2+}$ is blue in colour, because it absorbs red colour needed for excitation from the visible light, leaving the complementary blue coloured wavelenghts to pass on.

The colour exhibited depends upon the excitation state of the transition metal, *e.g.*,

Cr^{4+} (*blue*) ; Cr^{3+} (*violet*)
Mn^{2+} (*violet*) ; Mn^{3+} (*pink*)
Fe^{2+} (*green*) ; Fe^{3+} (*yellow*)

The non-metallic part associated with the transition metal also modifies the colour, *e.g.*, CuO (black), $Cu(OH_2)$ (pale blue), $CuCl_2$ (yellow), $CuBr_2$ (black), $CuSO_4.5H_2O$ (blue), etc.

Q. 12 : What is the blue complex ion formed, when a cupric salt solution in water is treated with ammonia ?

Ans. The blue complex ion so-formed in tetrammine copper (II) ion, $[Cu(NH_3)_4]^{2+}$.

Q. 13 : Explain why transition elements for alloys with iron.

Ans. An alloy is, usually, composed of two or more metals. *Since d-block elements possess small, but comparable size. so they can easily be mutually substituted by one another in crystal lattices.* Such a substitution results in the formation of *solid solution*—a class of alloys. Since iron is a transition element, so it can be substituted, in its crystal lattice, by any other transition metal. For example, nickel dissolves in iron to form a solution, in which nickel atoms replace iron atoms in the crystal lattice of iron. Hence, transition metals easily form alloys with iron.

Q. 14 : Explain why transition metals act an catalysts.

Ans. Explanation of catalytic powder of transition metals : The catalytic power of transition metals is probably through the use of the $(n-1)$ *d*-orbitals or due to the formation of some interstitial compounds, which adsorb and activate the reacting substances. Thus, they act as catalyst due to variation in oxidations states. Because of presence of *incomplete d-orbitals.* they can form unstable intermediate products with the reactants. *Transition metal compounds are believed to operate as catalyst by providing an alternative low energy pathways, either by variation in the oxidation state of transition metal or by the formation of suitable intermediates.* Thus, during the conversion SO_2 to SO_3 the catalyst V_2O_5 is believed to act due to the ability of vanadium to have several oxidation states. The probable mechanism for this catalytic action is as follows :

Solid V_2O_5 adsorbs a SO_2 molecule on its surface and gives it an oxygen atom to form divanadium tetroxide. The divanadium tetroxide is then converted back to V_2O_5 by the reaction with oxygen.

$$V_2O_5 + SO_2 \rightarrow V_2O_4 + SO_3$$
$$2V_2O_4 + O_2 \rightarrow 2V_2O_5$$

Q. 15 : Second ionization energies of chromium and copper are exceptionally higher than that of their neighbouring elements. Explain why.

Ans. The electronic configuration of :

$Cr+ = [Ar]\ 3d^5$

and $Cu+ = [Ar]\ 3d^{10}$.

Thus both Cr^+ and Cu^+ have stable cnfiguration of exactly half-filled [$3d^5$] and fully-filled [$3d^{10}$] *d*-orbitals. Consequently, removal of one electron form these, to give Cr^{2+} and Cu^{2+} ions, means change from a more stable state to less stable configuration, [Ar] $3d^4$ and [Ar] $3d^9$ respectively. Since such a change is quitc difficult, so second ionization energies of chromium and copper are sufficiently higher than those of their neighbours on the left as well as right.

Q. 16 : The first ionization energy of copper is higher than those of alkali metals. Explain why.

Ans. The electronic configuration of copper is [Ar] $3d^{10}$, $4s^1$ and removal of one electron results in the formation of Cu^+ with a stable electronic configuration of [Ar] $3d^{10}$ with full-filled $3d$-orbitals.

The electronic configuration of alkali metals in ns^1. *The removal of an electron from an alkali metal is relatively easier than from copper, because of larger atomic size of alkali metals than copper.* The attractive forces on the outer electrons in case of an alkali metal is lesser than that in copper. Hence, first ionization energy of copper is higher than those of alkali metals.

Q. 17 : Why *d*-block elements are metals, exhibiting good mechanical properties ?

Ans. *d-block elements are all metals.* They exhibit good mechanical properties such as hardness. malleability and ductility. This is because the number of valency electrons is fewer as compared to the available valency shell orbitals.

Q. 18 : Why do ionic radii decrease in order Ti^{2+}, V^{2+}, Cr^{2+},

Mn^{2+}, Fe^{2+}, Co^{2+}, Ni^{2+} and Cu^{+2}.

Ans. Ionic radii of the *d*-block elements follows the following trend :

$Ti^{2+} > V^{2+} > Cr^{2+} > Mn^{2+} > Fe^{2+} > Co^{2+} > Ni^{2+} > Cu^{2+}$.

Reasons : With the progressive increase in the atomic number, the nuclear charge increases, which pulls the electron-cloud inwards, thereby resulting in decrease is volume.

Q. 19 : Why are transition metal fluorides ionic in nature ; whereas bromides and chloride are covalent in nature ?

Ans. Transition metal fluorides are ionic in nature, because the difference in clectronegativities of transition metal and fluorine is quite large. On the other hand, the difference in electronegativities of transition metal and bromine (or chlorine) atom is less ; and hence, transition metal bromides and chlorides are covalent in nature.

Q. 20 : Explain the following :

(i) Why do Zr and Hf exhibit similar properties ?

(ii) Why is the separation of lanthanide elements difficult ?

(*All India Board, Outside Delhi, 1991*)

Ans. (i) Zr and Hf atoms are nearly *identical in size* (atomic radii of Zr and Hf are resectively 160 pm and 159 pm) and their outermost electronic configurations are also identical, so they exhibit similar properties.

(ii) Lanthanides are highly electropositive metals with similar outer electronic configurations, so they exhibit so much similar properties that their separation from the naturally occurring mixture is very difficult.

9

Coordination Compounds and Organometallics

Q. 1 : It is true that a cyclic complex is, usually, more stable than an open one. Substantiate your answer with *an* example.

Ans. Stability of chelates : Cyclic or chelate complexes are more stable than open complexes (*i.e.*, in which the ligand is a monodentate). For example cyclic like chlorophyll (a complex of Mg) and haemoglobin (a complex of Fe) are stable towards dissociation; while open complexes like $Ag(NH_3)_2^+$ and $Cu(NH_3)_4^{2+}$ are comparatively much less stable and get dissociated easily.

Q. 2 : Name the following co-ordination compounds using IUPAC norms :

(a) $[Cr(NH_3)_6]^{3+}$ (b) $[Ni(NH_3)_6]Cl_2$

(c) $[Mn(H_2O)_6]^{2+}$ (d) $[Co(en)_3]^{3+}$

(e) $[Fe(CN)_6]^{4-}$.

Ans.

Formula of Complex	*IUPAC Name*
(a) $[Cr(NH_3)_6]^{3+}$	Hexa-ammine chromium (III) ion
(b) $[Ni(NH_3)_6]Cl_2$	Hexa-ammine nickel (II) chloride
(c) $(Mn(H_2O)_6]^{2+}$	Hexa-aquo manganese (II) ion
(d) $[Co(en)_3]^{3+}$	Tris (ethylene diamine) cobalt (III) ion
(e) $[Fe(CN)_6]^{4-}$	Hexa-cyano ferrate (II) ion

Q. 3 : Write down the formulae of the following co-ordinate compounds :

(a) hexammine platinum (IV) chloride,
(b) potassium hexacyano ferrate (III),
(c) dichloro diammine platinum (II),
(d) dicholor tetrammine cobalt (III) ion,
(e) chloro-bis (ethylene diammine) ammine cobalt (III) ion.

Ans.

	IUPAC name of complex	*Formula*
(a)	Hexa-ammine platinum (IV) chloride	$[P (NH_3)_6]Cl_4$
(b)	Potassium hexacyano ferrate (III)	$K_3[Fe(CN)_6]$
(c)	Dichloro diammine platinum (II)	$[Po(NH_3)_2Cl_2]$
(d)	Dichloro tetra ammine cobalt (III) ion	$[Co(NH_3)_4Cl_2]^+$
(e)	Chloro-bis (ethylene diammine) ammine cobalt (III) ion.	$[Co(en)_2Cl.NH_3]^{2+}$

Q. 4 : Give the names of the following co-ordination compounds:

(i) $K_3[Fe(C_2O_4)_3]$.
(ii) $[CoCl(NH_3)_5Cl_2$.
(iii) $K_3[Fe(CN_6]$
(iv) $Co(NH_3)_5Br]SO_4$.
(v) $[Co(NH_3)_5SO_4]Br$.
(vi) $[Co(NH_3)_5ONO]Cl_2$.
(vii) $[Cu(en)_2]SO_4$.
(viii) $[Co(NH_3)_4H_2O.Cl]Cl_2$.
(ix) $Na_3[ReO_2(CN)_4]$.
(x) $[Cu(NH_3)_4]_3.[CrCl_6]_2$.
(xi) $[Co(en)_2(NH_3)_2]Cl_2$.

Ans.

(i) Potassium trisoxalato ferrate (III).
(ii) Chloro penta-ammine cobalt (III) chloride.
(iii) Postassium hexacyano ferrate (III).
(iv) Bromo penta-ammine cobalt (III) sulphate.
(v) Sulphato penta-ammine cobalt (III) bromide.
(vi) Nitrito penta-ammine cobalt (III) chloride.

(vii) Bis (ethylene diammine) copper (II) sulphate.
(viii) Chloro aquo tatra-ammine cobalt (III) chloride.
(ix) Sodium dioxo tetra-cyano rhenate (V).
(x) Tetra ammine copper (II) hexachloro chromate (III).
(xi) Diammine bids (ethylene diammine) cobalt (II) chloride.

Q. 5 : How would you account for the following :

$[Ti(H_2O_6]^{2+}$ is coloured; while $[Sc(H_2O_6]^{3+}$ is colourless.

Ans. $[Ti(H_2O_6]^{3+}$ ion is *coloured* (purple), because it contains unpaired *d*-electrons, which can undergo *d-d* transition by absorbing energy corresponding to green and yellow components from visible light and the complimentary colours (a mixture of red and violet) is reflected. Hence, $[Ti(H_2O_6]^{3+}$ ion is purple coloured. On the other hand, $[Sc(H_2O_6]^{3+}$ ion is colourless, because its sturcture does not contain any unpaired electron in *d*-orbitals. Consequently, it cannot undego and *d-d* transition. Hence, $[Sc(H_2O_6]^{3+}$ ion is *colourless.*

Q. 6 : Write all the isomers of :

$[Co(NH_3)_6] [Cr(C_2O_4)_3]$

Ans. $[Co(NH_3)_6] [Cr(C_2O_4)_3]$ is an example of compound, which cxhibits ***co-ordination isomerism* in which both cation and anion are complexes and the ligands can be exchange between the complex cation and complex anion.** The isomers are:

(i) $[Co(NH_3)_6] [Cr(C_2O_4)_3]$
Hexammine cobaltate (III)
tris (oxalato) chromate (III)

(ii) $[Cr(NH_3)_6] [Cr(C_2O_4)_3]$
Hexammine chromium (III)
tris (oxalato cobaltate (III)

Q. 7 : Write all the isomers of $[Co(NH_3)_5SCN]Cl_2$.

Ans. Isomers of $[Co(NH_3)_5SCN]Cl_2$ are :

(i) $[Co(NH_3)_5SCN]Cl_2$, and
(ii) $[Co(NH_3)_5CNS]Cl_2$.

Q. 8 : Write all isomers of : $[Pt(SCN) NH_3)_3]$ (SCN).

Ans. The isomers are :

[Pt(SCN) $NH_3)_3$] (SCN)

and [Pt(CNS) $NH_3)_3$] (SCN).

Q. 9 : Define an organometallic compound.

Ans. Organometallic compounds ***are those which contain one or more metal-carbon bonds.*** Usually, metalloids (such as germanium and antimony), metals such as aluminium, zine, lead, tin, magnesium, cobalt, iron, nickel, etc.) and non-metals (such as boron and silicon, etc.) are found to form organometallic compounds.

Examples : (i) Grignard's reagent.

$$R - Mg - X \quad \text{where} \begin{cases} R = \text{alkyl or aryl group} \\ X = \text{halogen atom} \end{cases}$$

(ii) Lead tetra ethyl, Pb $(C_2H_5)_4$.

(iii) Tin tetramethy1, $(CH_3)_4Sn$.

(iv) Zinc diethy1, $(C_2H_5)_2Zn$.

(v) Metal carbonyl, $M(CO)_n$ *e.g.*,

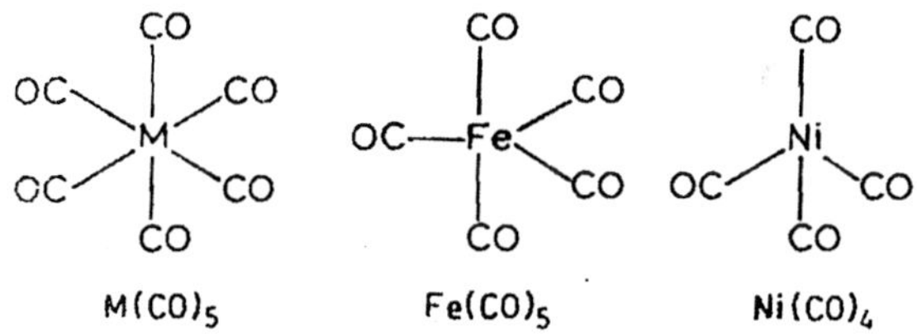

(vi) Dibenzene chromium. Cr $[\eta^6\text{–}C_6H_6]_2$.

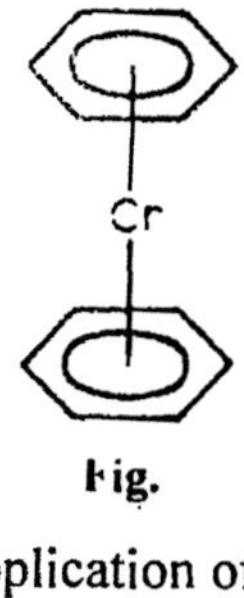

Fig.

Q. 10 : Mention the application of organometallic compounds in the following areas :

(a) Homogeneous catalysis.

(b) Heterogeneous catalysis.

(c) Organic synthesis.

Ans. (a) **Homogeneous catalysis :** Organometallic compounds derived from solube transition metal complexes are found to catalyse a number of reaction is solution. This is known as *homogeneous catalysis.* For example, in *Wilkinson's synthesis,* organometallic compound, $(Rh_3P)_3RhCl$ is used as a homogeneous catalyst during the hydrogenation of alkenes.

(b) **Heterogeneous catalysis :** Organometallic compound can act as heterogeneous catalysis. For example, Ziegler-Natta catalysts (organometallic compounds) are used as heterogeneous catalysts, during polymerization of olefins.

(c) **Organic synthesis :** Organometallic compounds are used to carry but a wide variety of organic synthesis. For example, alkyl magnesium halides Grignard's reagents) are used to synthesize' a large number of organic compounds.

10

Symmetry Properties

Q. 1 : Tridymite, the high temperature form of SiO_2, is hexagonal with a = 0.503 nm c = 0.822 nm. The space group is P6/mmc, and there are 4 SiO_2 units per unit cell. The Si atoms occupy positions with C_{3v} symmetry :

$$\frac{1}{3}, \frac{2}{3}, z; \quad \frac{2}{3}, \frac{1}{3}, \bar{z}; \quad \frac{2}{3}, \frac{1}{3}, \frac{1}{2}+z; \quad \frac{1}{3}, \frac{2}{3}, \frac{1}{2}-z$$

with z = 0.44.

The oxygen atoms occupy positions with $D_{3h} - \bar{6}m2$ symmetry :

$$\frac{1}{3}, \frac{2}{3}, \frac{1}{4}; \frac{2}{3}, \frac{1}{3}, \frac{3}{4}$$

The other six oxygens are in positions of $C_{2h} - 2/m$ symmetry :

$$\frac{1}{2}, 0, 0; \ 0, \frac{1}{2}, 0; \ \frac{1}{2}, \frac{1}{2}, 0; \ \frac{1}{2}, 0, \frac{1}{2}; \ 0, \frac{1}{2}, \frac{1}{2}; \ \frac{1}{2}, \frac{1}{2}, \frac{1}{2}$$

(a) Calculate the density of tridymite.

(b) Draw the projection of this structure on the ab plane, showing all of the atoms in at least one unit cell.

(c) Calculate the distance from a silicon atom to each of the two types of oxygen bonded to it.

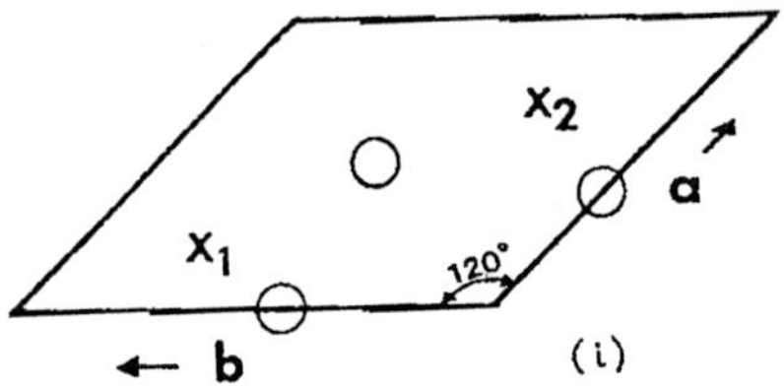

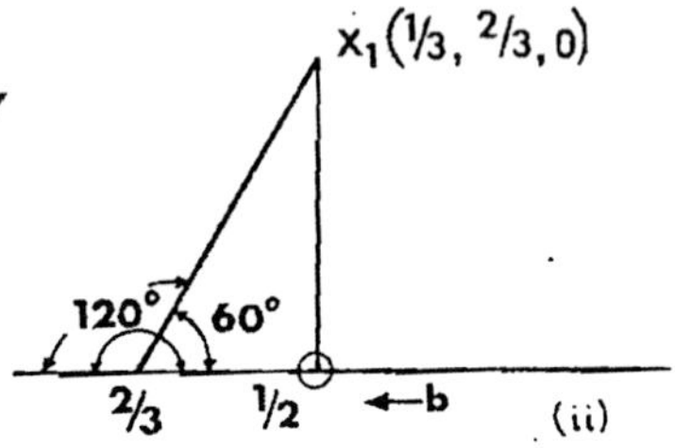

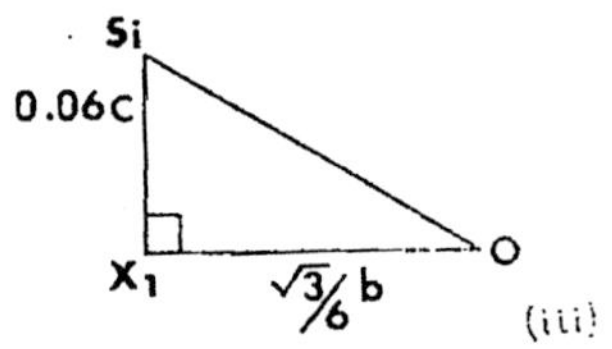

Solution : The volume of a unit cell is given by

$$V = abc\left(1 - \cos^2\alpha - \cos^2\beta - \cos^2\gamma + 2\cos\alpha\ \cos\beta\ \cos\gamma\right)^{\frac{1}{2}}. \quad (1)$$

Where V = volume

a, b and c = dimensions

α, β and γ = angles between the edges.

For the particular cell described above, equation (1) reduces to

$$V = abc\left(1 - \cos^2\alpha\right)^{\frac{1}{2}} \quad (2)$$

with $\alpha = 120°$.

Using the appropriate trigonometric identity, equation (2) takes the form

$$V = abc\left(\sin^2\ \alpha\right)^{\frac{1}{2}}$$

$$= abc\ \sin\alpha$$

Because of the hexagonal form, a = 0.503, b = 0.503 and c = 0.822. Therefore, $V = (0.503)^2\ (0.822)\ \sin 120° = 0.180\ nm^3$

or $V = 180 \times 10^{-24}\ cm^3$.

The density of a unit cell is related to its dimensions and angles by the expression

$$d = \frac{CM}{AV}$$

where

d = density

C = the contents of the unit cell (or the number of points contained within it.)

M = Molecular weight.

A = Avogadro's number.

V = Volume.

C is given to be 4 and the molecular weight of SiO_2 is 28.08 + 32 = 60.08. Sustituting the respective values into equation (4) gives

$$d = \frac{(4)\ (60.08)}{(6.02 \times 10^{23})\ (180 \times 10^{-24})}$$

$$= 2.218\ g\ cm^{-3}.$$

(b) Above each C_{2h} –2/m, oxygen in the ab plane (represented by 0) is an oxygen at $\frac{1}{2}C$. At $x_1\left(\frac{1}{3}a,\ \frac{2}{3}b\right)$, D_{3h} - $\bar{6}m2$ oxygen is at $\frac{1}{4}C$ and Si is at 0.44C and 0.06C.

At $x_2\left(\frac{2}{3}a,\ \frac{1}{3}b\right)$, D_{3h} - $\bar{6}m2$ oxygen is at $\frac{3}{4}C$ and Si is at 0.56C and 0.94C.

(c) The shortest distance between them lies on the vertical to ab plane. Therefore, for x_1 vertical, distance

$$d = (0.44 - 0.25)\ C = 0.156\ Si\ - C_{2h} - 2/m\ oxygen.$$

For example, distance from $\left(0,\ \frac{1}{2},\ 0\right)$ oxygen to $\left(\frac{1}{3}, \frac{2}{3},\ 0.06\right)$ Si is given by the distance from oxygen to x_1 on ab plane: that is :

$$\cos\ 60° = \frac{\left(\frac{1}{3}a\right)^2 + \left(\frac{1}{6}b\right)^2 - (x_1 - 0)^2}{2\left(\frac{1}{3}a\right)\left(\frac{1}{6}b\right)}$$

Note that a = b. Solving for $(x_1 - 0)$ yields

$$(x_1 - 0) = \sqrt{\frac{3}{6}}b$$

Now $(si - 0)^2 = (0.06c)^2 + \frac{3a^2}{36}$ from for figure.

$$= (0.0493)^2 + (0.1454)^2$$

$$(Si - 0) = 0.1535 \text{ nm.}$$

Q. 2 : List and explain all the symmetry operations on molecules.

Solution: Symmetry (or lack of symmetry) is a property of all objects. Objects can be classified according to their symmetry by looking for all the operations that leave them looking the same. These are called symmetry operations. For instance, a sphere is said to be highly symmetrical because no matter how the sphere is rotated it is indistinguishable for its original state. A pencil is said to have less symmetry because although any rotation about its length will leave it identical to its original state, a rotation of 90° perpendicular to its lenght will not. This symmetry is shown in Figures 1 through 3.

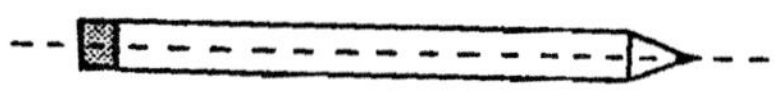

Fig. 1

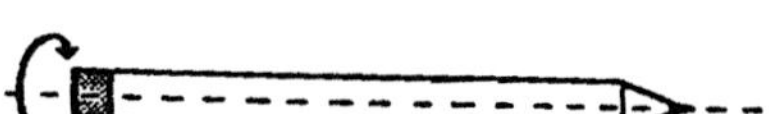

Fig. 2

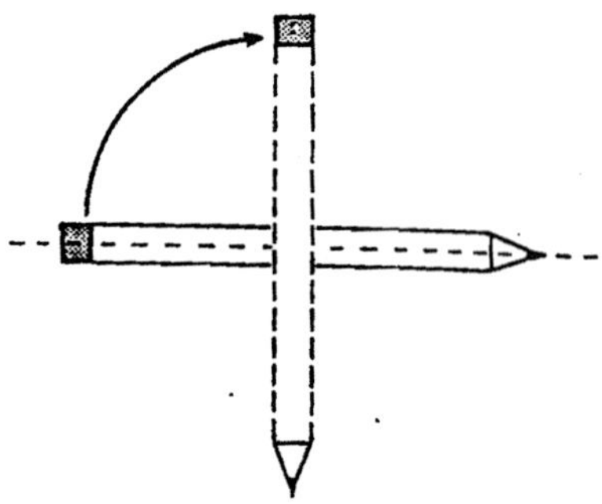

Fig. 1 Original.
Fig. 2 Rotation about its length
Fig. 3 Rotation of 90° perpendicular to its length

The same principle applies to molecules. For molecules, there are five symmetry operation that can be performed on them. They are identity, rotation about an axis of symmetry, reflection through a plane of symmetry (i.e., a mirror plane), inversion through a center of symmetry, and improper rotation. Crystals contain two more unique symmetry operations, translation and screw rotation. The use of these symmetry make it possible to classify all the different types of molecules.

(1) The identity operation performs no operation on the molecule. All molecules are identical to their orginal position if the identity operation is done. This may seem superfluous. However, it allows a molecule such as CHClBrF to be classified. This operation shall be denoted by E.

(2) Rotation about an axis of symmetry is that operation which allows a molecule to be rotated through an angle of 360°/n and remain unchanged from its original condition. The molecule is then said to have an n-fold axis of symmetry. This is denoted by the symbol C_n.

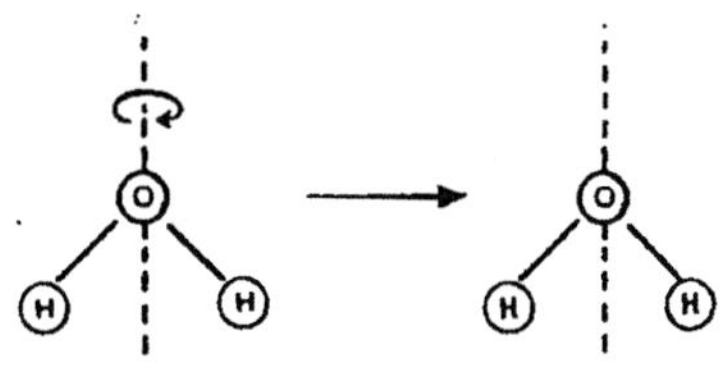

Fig. 4

For instance, if a water molecule is rotated by 180° it is indistinguishable from its original state. This is shown in Figure 4.

Hence, the water molecule has a C_2 or two-fold axis of symmetry (360/2 = 180). If there are several axes of symmetry, the one with the highest value of n is used. This is called the principal axis of symmetry.

(3) Reflection through a plane of symmetry (a mirror plane) is when reflection in a plane passing through the molecule leaves the molecule indistinguishable from its original state.

When the principal axis is contained in the mirror plane this symmetry element (operation) is denoted σ_v (the v denoting verical).

For instance, the water molecule shown in Figure (5) has a twofold axis of symmetry passing through the oxygen atom. In addition to the C_2 axis, there are two σ_v plances. Both contain the C_2.

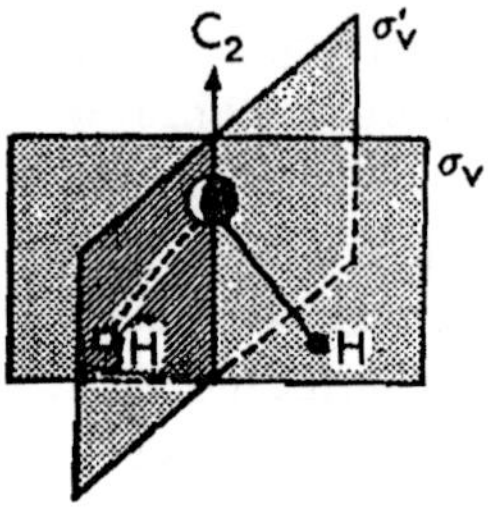

Fig. 5

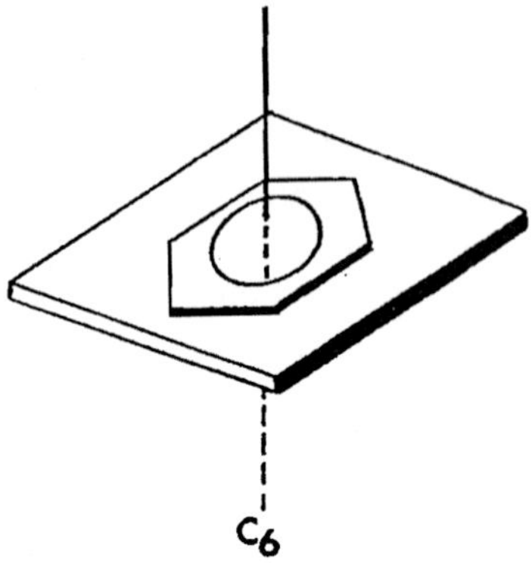

Fig. 6

axis, and one is just the H–O–H plane, while the other is perpendicular to it.

When the mirror plane is perpendicular to the principal axis (which if C_n is a vertical axis makes σ a horizontal plane), it is denoted by σ_h. An example of this is benzene, which is shown in Figure 6.

Thus benzene has a six-fold axis of symmetry. (The person observing this is looking down the axis of symmetry and seeing the benzene molecule lying flat).

There is one other σ plane, σ_d, which is much trickier and harder to see. Consider the hypothetical molecule shown in Figure 7.

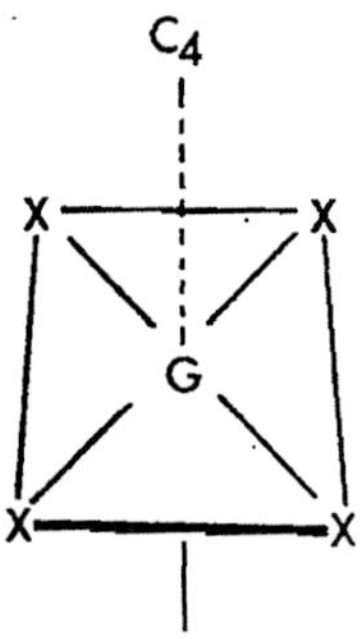

Fig. 7

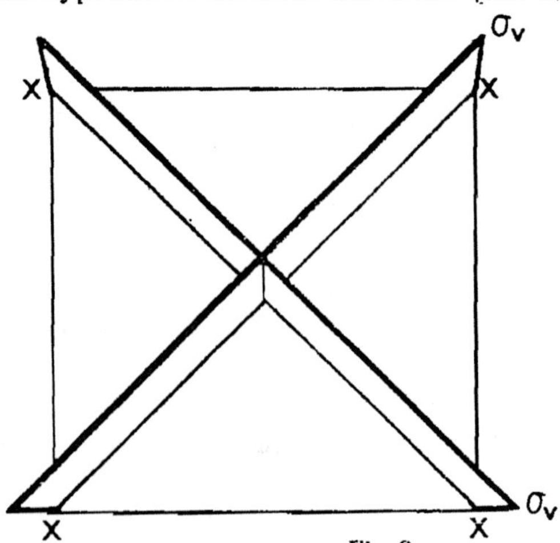

Fig. 8

where all the x's lie in a plane. One can find a C_4 axis of symmetry since rotation by 90° leaves the molecule unchanged. Now the σ_v planes can also be found. (The person looking at this figure is looking down the C_4 axis, which goes from above the plane of the page and extends below it, and sees the molecule flat and in the plane of the page). They are the ones that run along the diagonals, as seen in Figure 8.

There is also the horizontal σ_h plane shown in Figure 9.

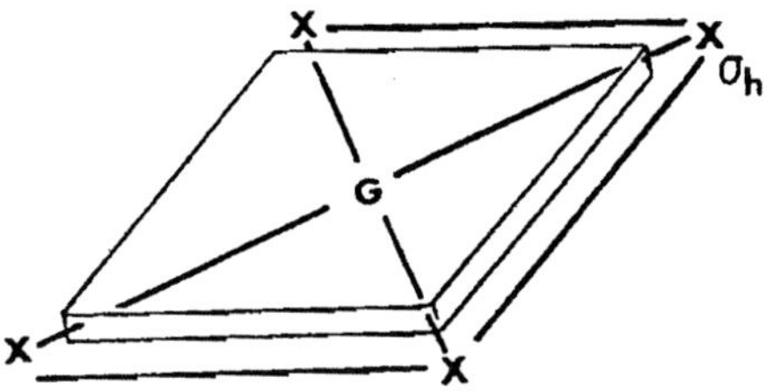

Fig. 9

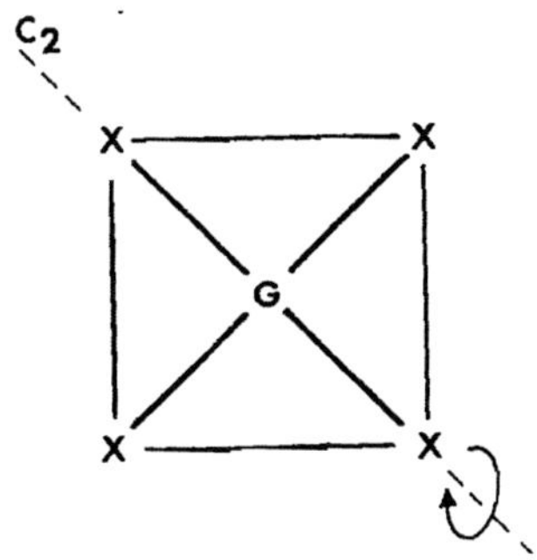

Fig. 10

Now, this molecule has other axes of symmetry besides the principal C_4 axis. The particular ones that are important for determining J_d are the two C_2 axes.

Consider the molecule in Figure 10. There is a C_2 axis running along the diagonal of the molecule in the plane of the paper. If the molecule is rotated 180° around this axis, it is indistinguishable from its starting position. Likewise, for the other C_2 axis. See Figure 11.

The two σ_d planes are those planes which are vertical and contain the principal axis and also bisect the angle between the two C_2 axes (which are themelves perpendicular to the principal axis).

The overall result of all the σ planes mentioned is Figure.12.

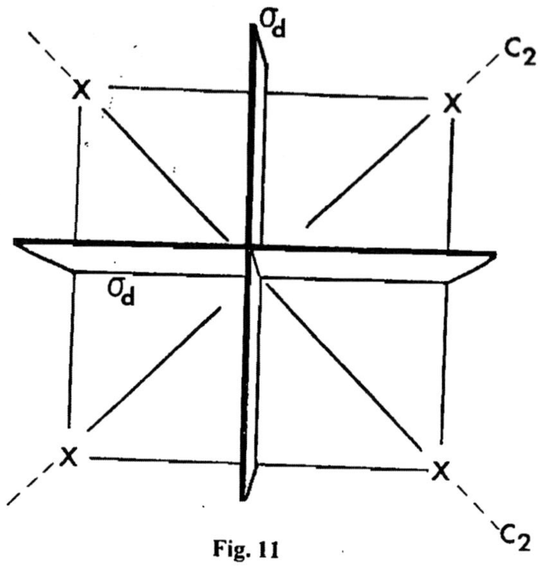

Fig. 11

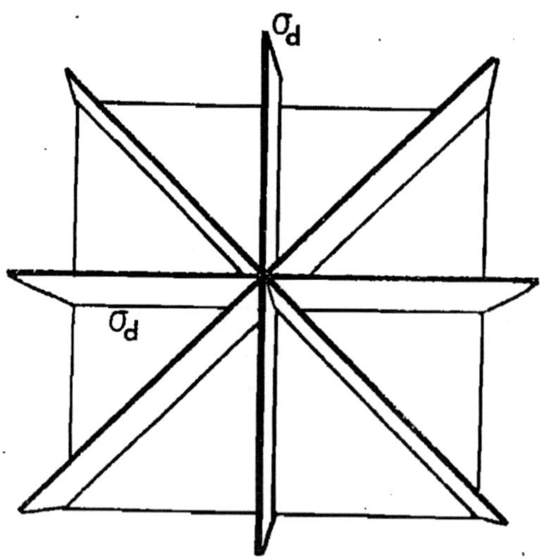

Fig. 12

(4) Inversion is when the molecule is left unchanged after all its points in space are inverted. In other words, a center for the molecule can be found where, if the distance out in one direction is reversed to the opposite direction, the molecule is unchanged. For instance, in Figure 13 (where a heavy line denotes coming out of the plane of the paper and a dashed line denotes going into the plane of the paper) there exists

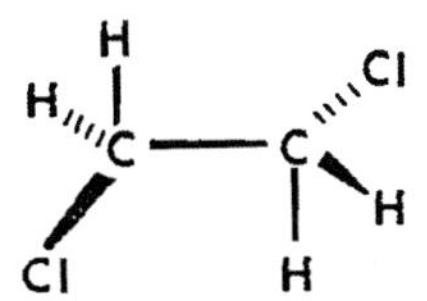

Fig. 13

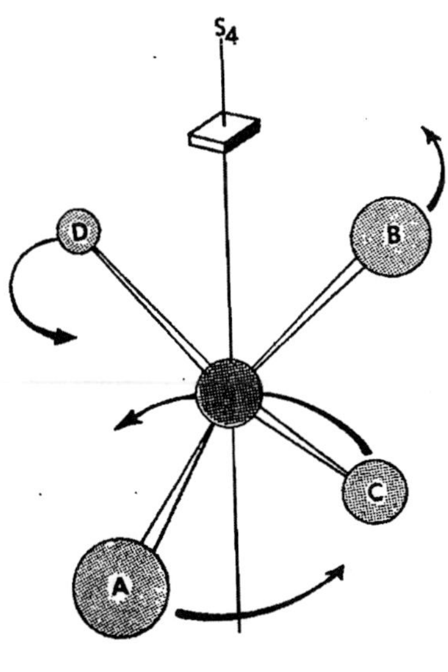

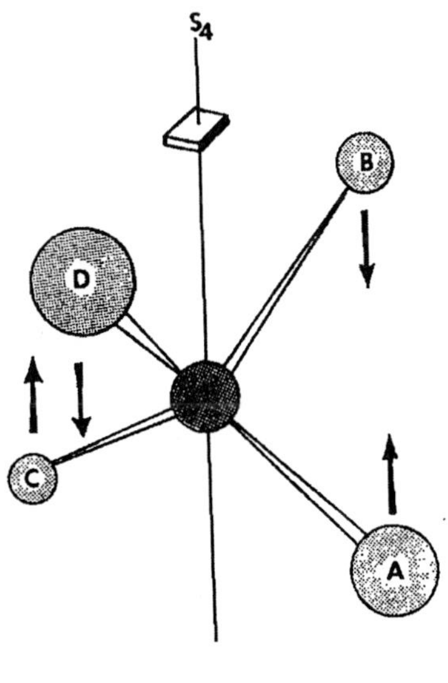

Fig. 14

Fig. 15

a center of symmetry so the chlorine coming out of the plane on the first carbon if inverted would coincide with the chlorine on the second carbon going into the plane. Like-wise, for all the other atoms. The operation is denoted by i. (5) Improper rotation (or rotary-reflection) is actually two operations. A molecule possesses it if it is unchanged after first under-going an n-fold rotation followed by a horizontal reflection. This operation is denoted by S_n. An example of this is shown in.

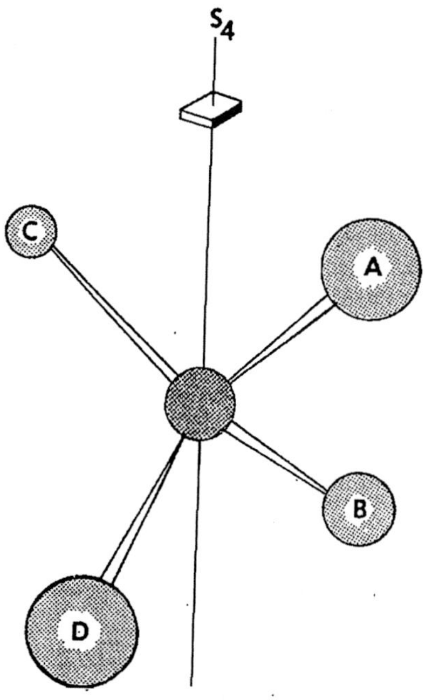

Fig. 16

Figure 14 through 16. The letters are really to aid the reader. For a real molecule such as CH_4 all the hydrogens are equivalent so one could not tell which one was A, B, C, or D. Hence, Figure 14 is equivalent to Figure 16.

Q. 3 : Explain how point groups for molecules are designated from the symmetry operations.

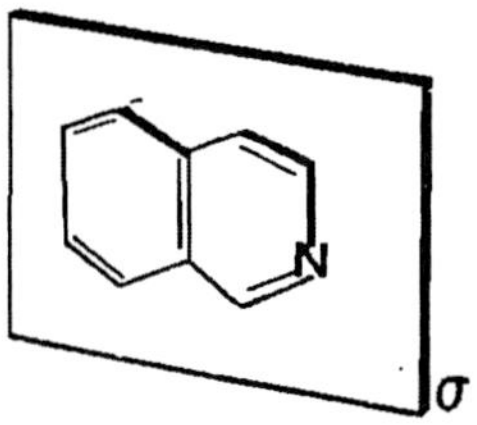

Fig. 1

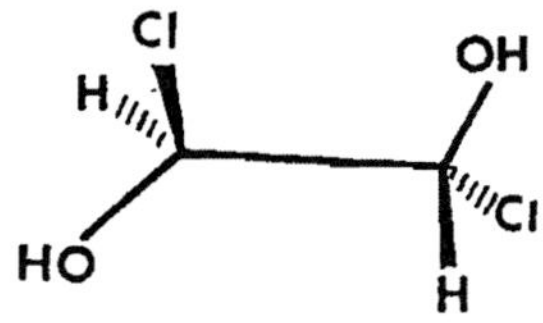

Fig. 2

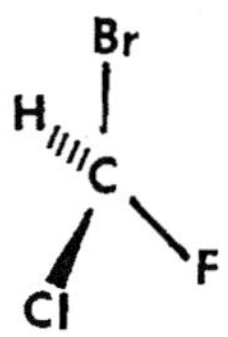

Fig. 3

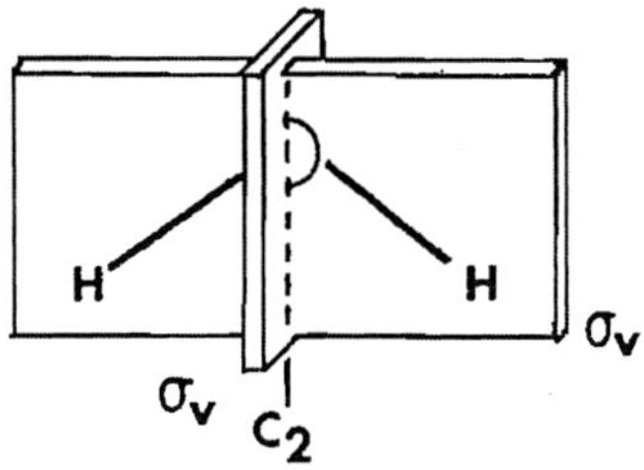

Fig. 4

Solution: Point groups are defind according to the type of symmetry elements they contain.

(1) The C_s group is that group where the molecule contains only the identity element and a plane of symmetry. An example of this would be isoquinoline, shown in Figure 1.

(2) The C_i group is that group where the molecule contains only the identity element and a center of inversion. An example of this is shown in Figure 2. Since all the other groups contain the identity operation it will not be explicitly mentioned for them.

(3) The C_n group is that group which contains only an n-fold axis of symmetry. An example would be the compound CHClBrF, shown in Figure 3, which contains a one-fold axis of symmetry (since rotation by 360° leaves the molecule unchanged and n = 360/360 = 1. Hence, CHClBrF is C_1.

(4) The C_{nv} group is that group which contains a C_n axis and n vertical reflection palnes σ_v. An example would be H_2O, shown in Figure 4. Hence, water is a member of the group C_{2v}.

(5) The C_{nh} group is that group where the molecule contains a C_n axis and a horizontal plane of reflection. An example of this would be trans CHCl = CHCl, shown in Figure 5.

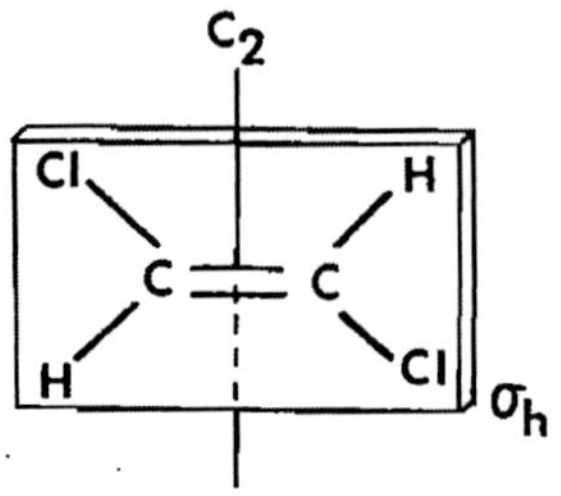

Fig. 5

This compound has a C_2 axis of symmetry and a σ_h plane of symmetry and hence it is a member of C_{2h}.

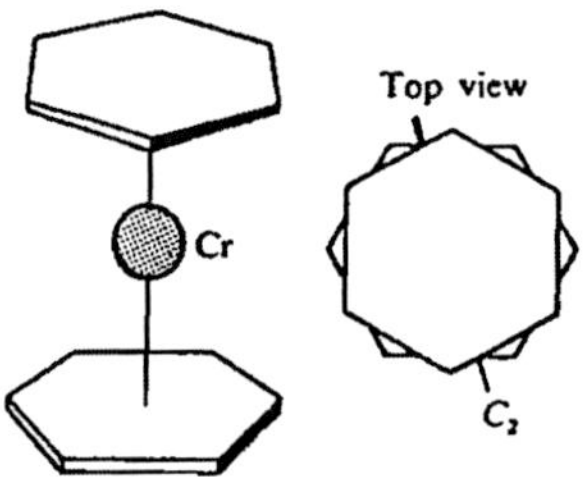

Fig. 6

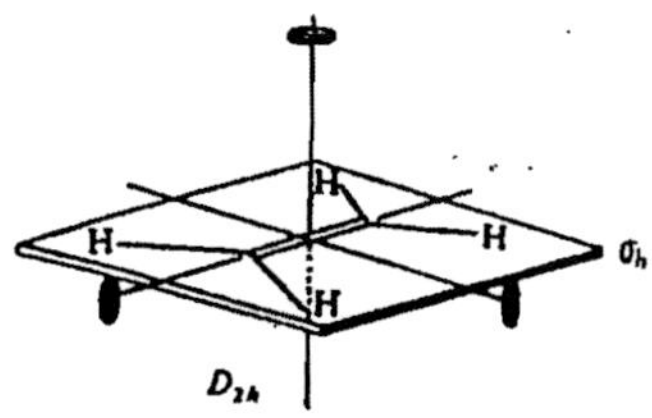

Fig. 7

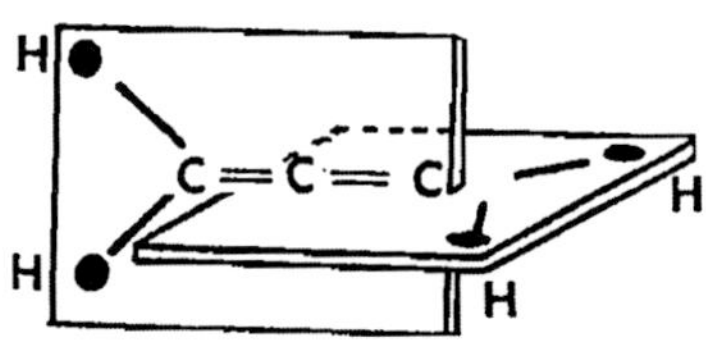

Fig. 8

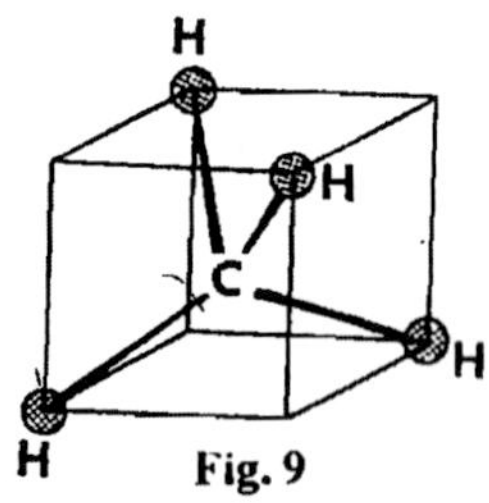

Fig. 9

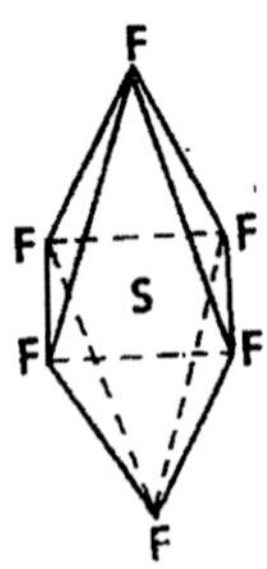

Fig. 10

(6) The D_n group is that group where the molecule possesses a C_n axis and an n-two fold axes perpendicular to C_n. An example is shown in Figure 6.

Thus, this molecule is a member of D_2.

(7) The D_{nh} group is that group where the molecules belong to D_n and possesses a horizontal plane of symmetry, σ_h. An example of this is ethylene, C_2H_4 shown is Figure 7.

(8) The D_{nd} group is that group where the molecules belong to D_n and possesses an additional vertical plane which bisects the angles between all the adjacent C_2 axes. An example of this is allene, shown in Figure 8 which is D_{2d}. The next groups to be considered have more than one axis of symmetry where the order of rotation is greater than two. They are the octahedral and tetrahedral groups.

(9) The T_d group is that group where the molecule has a regular tetrahedron shape. An example of this is methane, CH_4, shown in Figure 9. If the T_d group possesses in addition rotational symmetry, it becomes the group T. If the group T contains a center of inversion it becomes the group T_h.

(10) The O_h group is that group where the molecule has a regular octahedron. An example of this is SF_6, shown in figure 10. If the O_h group possesses rotational symmetry it becomes the O group.

Q. 4 : For each molecule below, determine its point group :
(a) N_2, (b) PCl_5, (c) NH_3, (d) C_2H_4, (e) pyridine, (f) pyrazine, (g) hydrogen peroxide, (h) cyclopropane (ignore the hydrogens), (i) C_2H_5Cl (in eclipsed form), (j) SF_5Cl (octahedral), (k) CO.

Fig. 1

Fig. 2

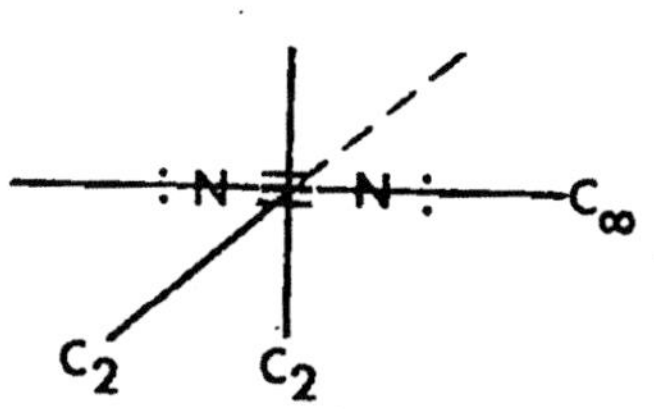

Fig. 3

Fig. 4

Solution: (a) N_2 is a linear homonuclear molecule. It has the following structure shown in Figure 1. One can draw an axis of symmetry right through the molecule as shown in Figure 2. No matter how small an angle θ this molecule is rotated about this axis, the molecule is unchanged. Hence, N_2 has a C_∞ axis of symmetry. As n approaches zero, 360/n approaches infinity. The molecule also possesses n 2-fold axes. An example of two of them is shown in Figure 3 where the vertical C_2 is in the plane of the paper and the other starts above the plane of the paper and extends below it. Hence, the molecule is a member of D_n. N_2 also possesses a horizontal reflection plane, σ_h, perpendicular to the principal axis as seen in Figure 4. Therefore, N_2 belongs to the group $D_{\infty h}$.

(b) PCl_5 has 3Cl atoms in a plane at 120° and two Cl atoms above and below the phosphorous atom 180° apart. The molecule is shown in Figure 5. The molecule contains a C_3 axis passing through the phosphorous atom and the two Cl atoms above and below it. It also possesses $3C_2$ axes along each P-Cl bond in the plane. Since the molecule contains a horizontal mirror plane, σ_h, the molecule belongs to the group D_{3h}. These features can be seen in Figure 6.

(c) NH_3 has a structure that can be seen in Figure 7. It has a C_3 axis of symmetry that starts above the nitrogen atom and extends to the middle of the triangular base. There are also 3 vertical mirror planes, σ_v which contain the principal axis. Each of these cut through the nitrogen atom and one of the hydrogens. This is shown in Figure 8. Hence, NH_3 is a member of the group C_{3v}.

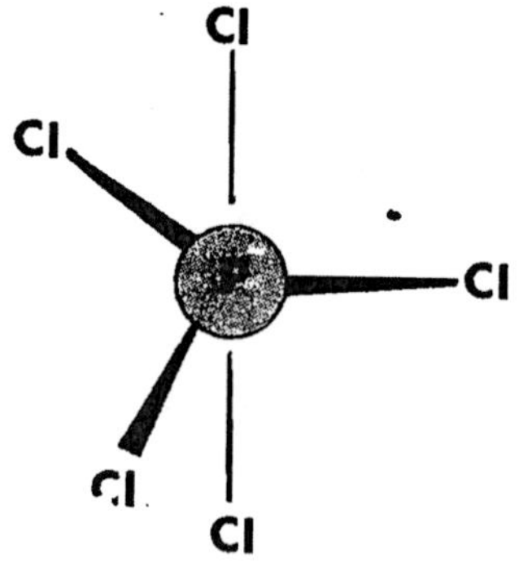

Fig. 5

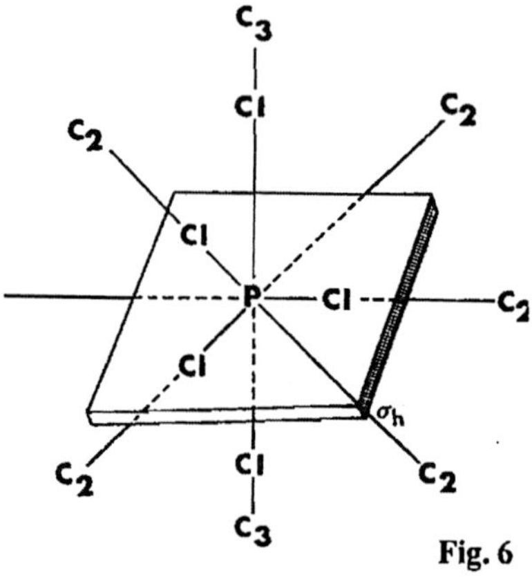

Fig. 6

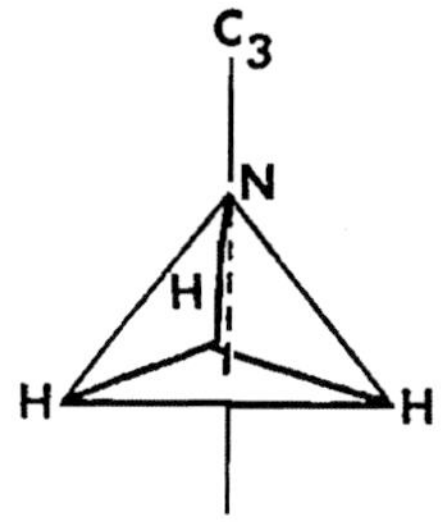

Fig. 7

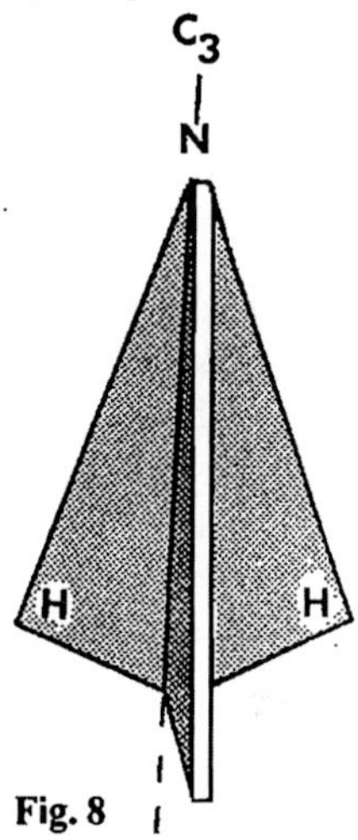

Fig. 8

(d) C_2H_4 is a planar hydrocarbon molecule. Its structure is shown in Figure 9. This molecule has a principal axis starting above the plane of the molecule and extending below it. Since rotation by 180°

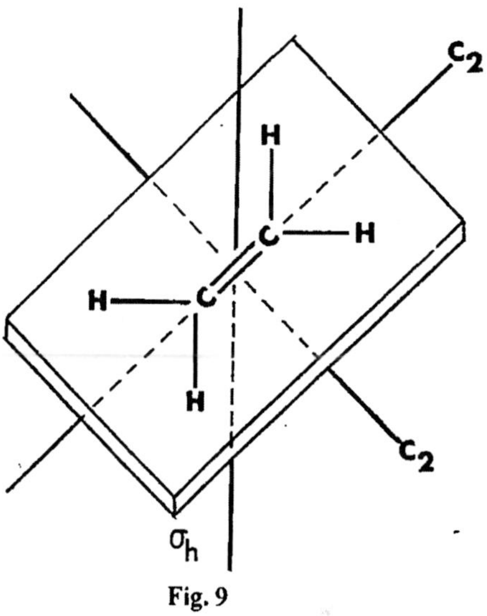

Fig. 9

Fig. 10

will leave the molecule unchanged, the principal axis is a C_2 axis. There are 2 C_2 axis perpendicular to the principal axis. They lie in the plane of the molecule. Hence, the molecule is a member of D_2. In addition, the molecule contains a horizontal mirror plane, σ_h, perpendicular to the principal axis. Therefore, the molecule belongs to the group D_{2h}.

(e) Pyridine is a heterocyclic aromatic compound. It has the following structure, as shown in Figure 10. A C_2 axis of symmetry can be found running through the plane of the molecule from the nitrogen atom to the opposite carbon atom. In addition, there are two vertical mirror planes σ_v, which contain the principal axis, hence pyridine is in the C_{2v} group. These features are shown in Figure 11.

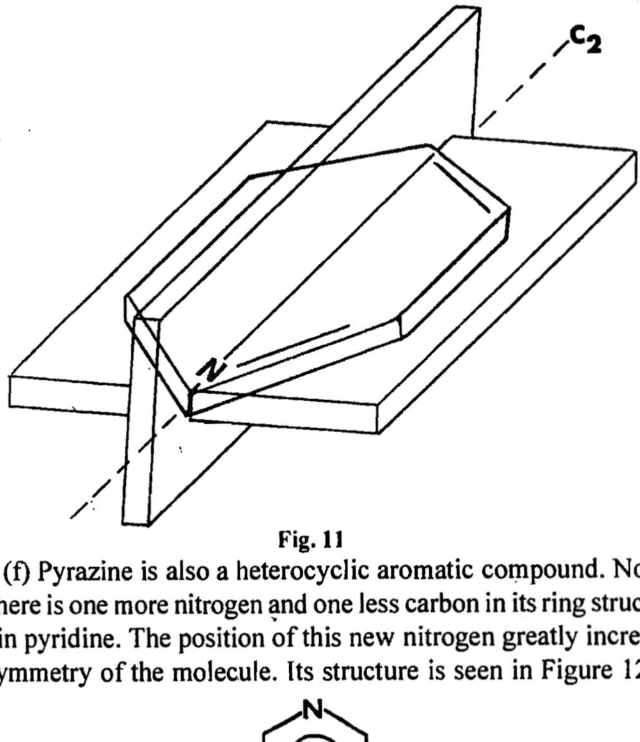

Fig. 11

(f) Pyrazine is also a heterocyclic aromatic compound. Notice that there is one more nitrogen and one less carbon in its ring structure than in pyridine. The position of this new nitrogen greatly increases the symmetry of the molecule. Its structure is seen in Figure 12. In

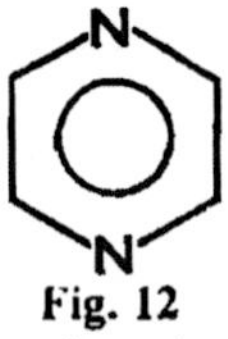

Fig. 12

this case, there is a C_2 axis starting above the plane of the molecule and extending below it. There are two C_2 axes perpendicular to the principal C_2 axis, which run through the plane of the molecule. In addition, there is a horizontal mirror plane running perpendicular to the principal C_2 axis. Hence, the pyrazine molecule belongs to the point group D_{2h}. These features are shown in Figure 13.

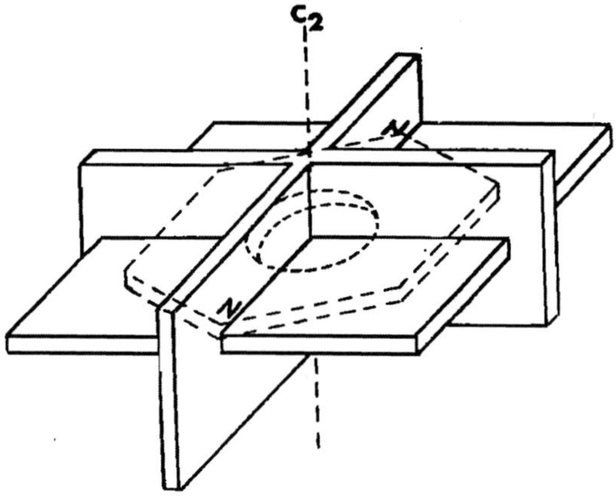

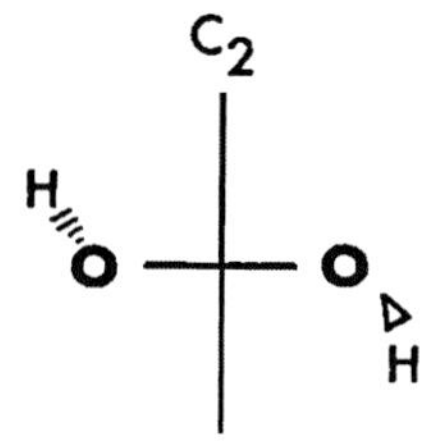

Fig. 13 Fig. 14

(g) Hydrogen peroxide, H_2O_2, has the following structure, shown in Figure 14. The only symmetry element this molecule is C_2. Rotating this molecule about this principal axis 180°, brings the hydrogen out front (on the right) to the back (on the left) and the hydrogen in the back (on the left) to the front (on the right). Hence, H_2O_2 is a member of the C_2 point group.

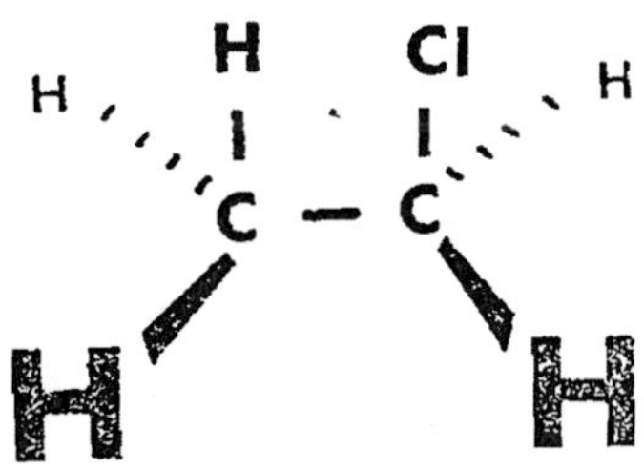

Fig. 15

(h) Cyclopropane without considering the hydrogens will be a planar, triangular shaped molecule. It will have a C_3 principal axis extending through the molecule from above it to below it. There will also be 3 C_2 axis perpendicular to the principal axis. They are in the plane of the molecule and run from one vertex to the opposite side. There is also a horizontal mirror plane perpendicular to the C_3 axis. These features indicate cyclopropane is of the point group D_{3h}, and they are shown in Figure. 16.

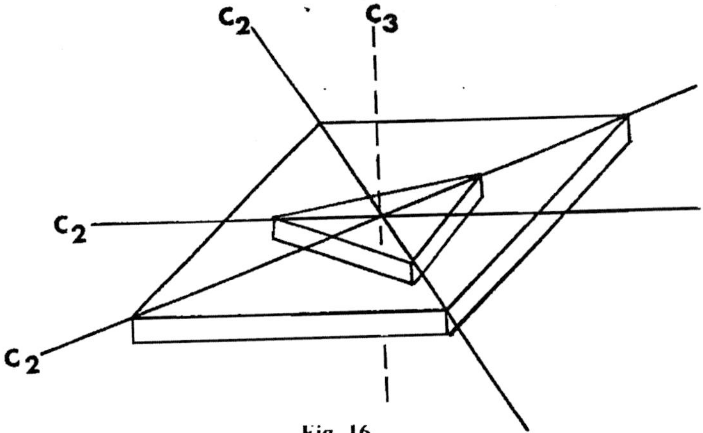

Fig. 16

(i) There is only one plane of symmetry in this molecule. This plane contains the 2 carbons and the H and the Cl pointing up. If this plane is inserted what is behind the plane is the same as what is in front of it. Since this is the only symmetry element present, this molecule belongs to the point group C_s. The diagram seen in Figure 17 shown this symmetry.

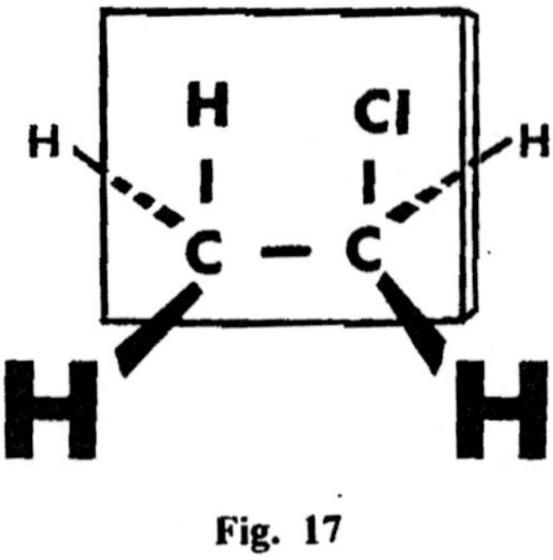

Fig. 17

(j) SF_5Cl (octahedral) has the structure shown in Figure 18. There is a C_4 axis running through the Cl–S–F atoms from above the plane to below it. There are also 4 vertical reflection plance which contain this C_4 axis. Two run between a pair of flourine atoms and the other two contain two flourine atoms along a diagonal. Figure 19 illustrates this (the F, Cl, and S have been left out for clarity). The point group for SF_5Cl is thus C_{4v}.

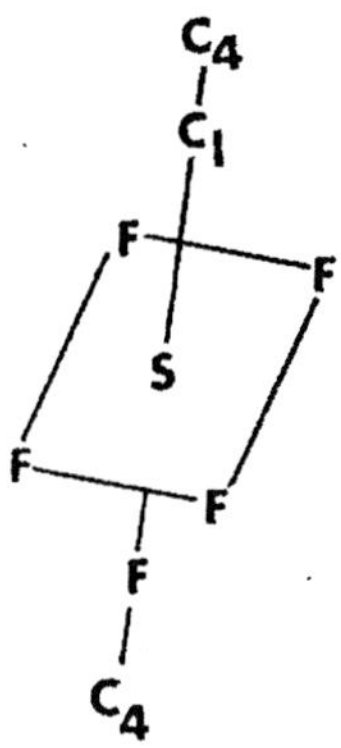

Fig. 18

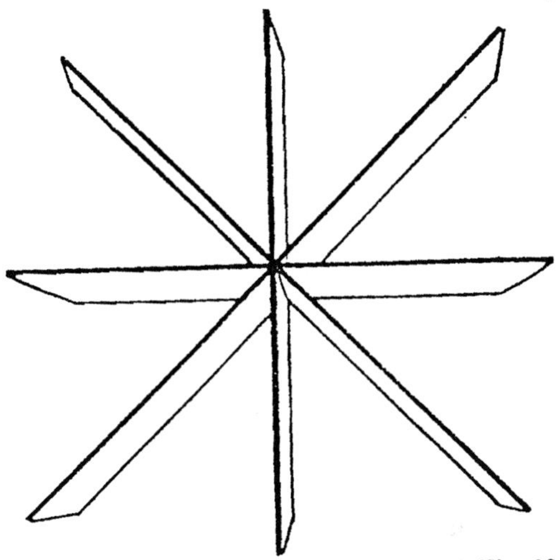

Fig. 19

Notice that if the chlorine atom were replaced by a flourine atom the symmetry would be greatly increased and the point group would be O_h.

(k) CO is a linear heteronuclear diatomic molecule. It has the structure shown in Figure 20.

:C≡O:

Fig. 20

One can draw an axis of symmetry right through the molecule as seen in Figure 21.

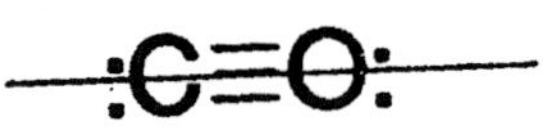

Fig. 21

No matter how small an angle θ this molecule is rotated about this axis, the molecule is unchanged. Hence, CO has a C_∞ axis of symmetry. As n approaches zero, 360/n approaches infinity. In addition, any vertical plane drawn through this molecule will contain the axis of symmetry (there will be an infinite number of these planes).

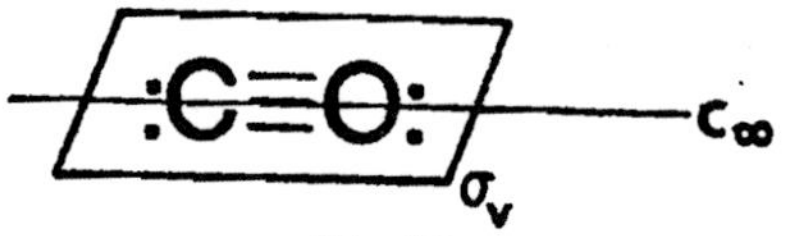

Fig. 22

This is shown in Figure 22. Hence, CO belongs to the point group $C_{\infty v}$.

11

Solid State Chemistry

Q. 1 : What is the plane of closest packing in the fcc and bcc structures ? What is the relation between the length of cube a_0 and the closest internuclear distance in these structures ?

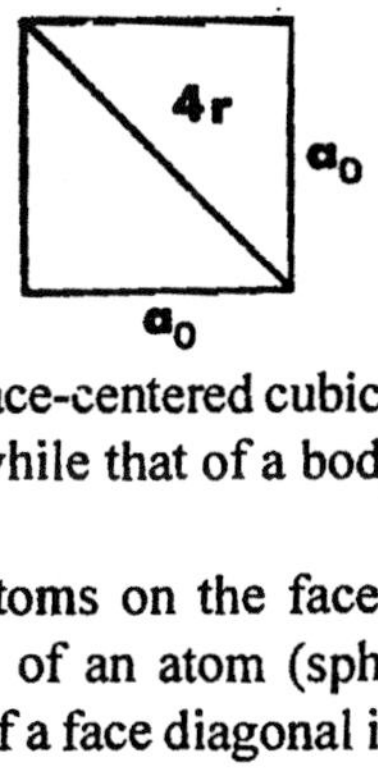

Solution: For a face-centered cubic unit cell, the plane of closest packing is 111 plane while that of a body-centered cubic unit cell is 110 plane.

There are two atoms on the face diagonal of an fcc. If r is taken to be the radius of an atom (spherical in shape) at a lattice point, then the length of a face diagonal is 4r. Note that there are 2r in each face diagonal.

From the diagram and using the pythagorean theorem,

$$a_0^2 + a_0^2 = (4r)^2$$

or $$2a_0^2 = (4r)^2$$

$\therefore$ $$a_0 = \frac{4r}{\sqrt{2}} = \frac{2(\sqrt{2})(\sqrt{2})r}{\sqrt{2}}$$

$$= 2\sqrt{2}r$$

Similarly, for a body-centered cube, the length of the body

diagonal is 4r; But this time, $(4r)^2 = a_0^2 + a_0^2 + a_0^2 = 3a_0^2$ from which

$$a_0^2 = \frac{(4r)^2}{3}$$

or $$a_0 = \frac{4r}{\sqrt{3}}$$

Q. 2 : Tetraborane is monoclinic with a = 0.868, b = 1.014, c = 0.578 nm, $\beta = 105.9°$. Draw a diagram showing the a and c axes and the orientation of the 101 planes. Calculate the interplanar spacing for 101 based on your diagram.

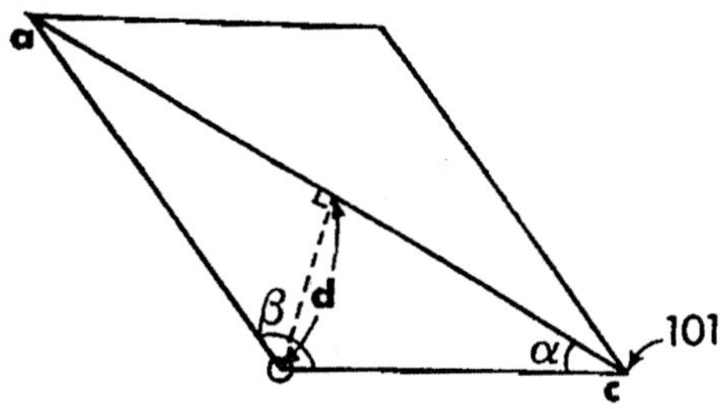

Fig.

Solution: The diagram that shown a and c axes and the orientation of the 101 plane is shown below as follows :

Let the distance between the axes a and c be h. By the cos ine rule,

$$h^2 = a^2 + c^2 - 2ac \cos \beta$$

or $$h = \sqrt{a^2 + c^2 - 2ac \cos \beta} \quad (1)$$

Substituting the given values into equation (1) gives

$$h = \sqrt{(0.868)^2 + (0.578)^2 - 2(0.868)(0.578) \cos 105.9°}$$

$$= \sqrt{0.753 + 0.334 - 1.003(-0.274)}$$

$$= \sqrt{1.087 + 0.2748} = \sqrt{1.3618}$$

$$= 1.167 \text{ nm.}$$

Similarly, $a^2 = h^2 + c^2 - 2hc \cos \alpha.$ (2)

From equation (2),

$$\cos\alpha = -\left(\frac{a^2 - h^2 - c^2}{2hc}\right)$$

$$= \frac{h^2 + c^2 - a^2}{2hc} \qquad (3)$$

Substituting for h, c and a in equation (3) yields

$$\cos\alpha = \frac{(1.167)^2 + (0.578)^2 - (0.868)^2}{2(1.167)(0.578)}$$

$$= \frac{1.362 + 0.334 - 0.753}{1.349}$$

$$= \frac{0.943}{1.349} = 0.699$$

or $\qquad \alpha = \cos^{-1} 0.699 = 45.65°$

Note that the desired distance is d. Using the relation

$$\sin\alpha = \frac{d}{c}$$

then $\qquad d = c\sin\alpha = (0.578)\sin 45.65°$

$$= 0.413 \text{ nm.}$$

Q. 3 : Sodium crystallizes in the body-centered cubic structure with a = 4.24 Å. Calculate the theoretical density of Na.

Solution : The content of a body-centered unit cell is 8(1/8) + 1 = 2. In a body-centered cell, there are eight corners and their contribution is 8(1/8) = 1. In other words, every corner of the unit cell contributes 1/8 of an atom. Also there is an atom on the body, which is not shared, making it 2 altogether.

$$d = \text{density} = \frac{ZM}{AV} \qquad (1)$$

Where Z = content = 2

M = Molecular weight

A = Avogadro's number

V = abc $(1 - \cos^2\alpha - \cos^2\beta - \cos^2\gamma + 2\cos\alpha\cos\beta\cos\gamma)^{1/2}$

But for a cell having 90° (as in this case), V = abc. a, b and c are lengths of the edges. V = Volume. Therefore

$$V = (4.24\text{Å})^3 = (4.24\times10^{-10}\,\text{m})^3$$

$$\text{Using}(1), d = \frac{(2)(23\text{gm mol}^{-1})(10^{-3}\,\text{kg gm}^{-1})}{(6.022\times10^{23}\,\text{mol}^{-1})(4.24\times10^{-10}\,\text{m})^3}$$

$$= 1.00\times10^3\,\text{kg m}^{-3}.$$

Q. 4 : Using the following ionic radii, find the fractional void volume in (a) CsCl, (b) NaCl, (c) LiCl. Assume that the ions are hard spheres in contact.

Ion	Radius, Å
Li^+	0.60
Na^+	0.95
Cs^+	1.69
Cl^-	1.81

In CsCl the positive ions and the negative ions each comprise a simple cubic lattice. Each ion has eight oppositely charged ions as nearest neighbors. In NaCl and LiCl the ions of each sign comprise a face-centered cubic lattice. Each ion has six oppositely charged ion as nearest neighbors.

Solution: (a) CsCl :

Since the CsCl is to be taken as a simple cubic lattice, the unit cell has two atoms. If the Cl^- ions are considered to be at the corners of the unit cell and the Cs^+ ion at the body, then the unit cell contains

$$1\ Cs^+ + 8(1/8)\ Cl^- = 2.$$

There are two possible ways to compute the length of the edge. First if the Cl^- ions are in contact along the edge, the length will be 1.81 + 1.81 = 3.62 Å. Second, if the Cs^+ and Cl^- ions are in contact along the body diagonal, then the length of the diagonal is

$$1.81 + 2\,(1.69) + 1.81 = 7.00\ \text{Å}.$$

Consequently, the length of the edge is

$$\frac{7.00}{\sqrt{3}} = 4.04\text{Å}.$$

Since the latter value is larger, the length of the edge must be

4.04 Å.

By definition, the volume of the unit cell is given by $V = a^3$.

Therefore, $V = (4.04\ \text{Å})^3 = 65.9\ \text{Å}^3$.

The ions are assumed to be spherical in shape and the volume of a sphere is given by $4/3\ \pi\ r^3$. For both Cs^+ and Cl^-, the volume is

$$\frac{4}{3}\pi(1.69)^3 + \frac{4}{3}\pi(1.81)^3 = \frac{4}{3}\pi(1.69^3 + 1.81^3)$$

$$= 4.2(4.82 + 5.9) = 45.1\text{Å}.$$

The fraction void is

$$\frac{65.9 - 45.1}{65.9} = \frac{65.9}{65.9} - \frac{45.1}{65.9}$$
$$= 1 - 0.684 = 0.316$$

(b) NaCl :

Since this is a face-centered cubic lattice, it has 4 "atoms". Therefore, it contains 4 Na^+ + 4 Cl–. If the Cl^- ions are considered to be in contact along a face diagonal, then the diagonal is

$$4r - (4)(1.81) = 7.24\ \text{Å long.}$$

As a result, the length of the edge is

$$\frac{7.24}{\sqrt{2}} = 5.12\text{Å}.$$

Secondly, if Na+ and Cl^- are in contact along an edge, then the edge is

$$2\ (0.95 + 1.81) = 5.52\ \text{Å}.$$

If on the other hand the Na^+ and Cl^- are considered to be in contact along the body diagonal, this diagonal is

$$2\ (0.95 + 18.1) = 5.52\ \text{Å},$$

and the edge is

$$\frac{5.52}{\sqrt{3}} = 3.19\text{Å long.}$$

The largest of these values must be the correct one. Consequently, the length of the edge is 5.52 Å.

The volume of the unit cell is therefore

$$(5.52\ \text{Å})^3 = 168\ \text{Å}^3.$$

The Voluem of

$$4Na^{+} = 4\left(\frac{4}{3}\pi(0.95)^{3}\right) = 14.37Å^{3}$$

while that of

$$4Cl^{-} = 4\left(\frac{4}{3}\pi(1.81)^{3}\right) = 99.37Å^{3}.$$

The total volume of the ions is

$$14.37 + 99.37 = 113.74 \text{ Å}^{3}.$$

The fraction void volume

$$= \frac{168 - 113.74}{168}$$

$$= 1 - \frac{113.74}{168} = 0.32.$$

(c) LiCl :

For this, edge contact is 2 (0.60 + 1.81) 4.82 Å. Face diagonal contact is

$$4\left(\frac{1.81}{\sqrt{2}}\right) = 5.12Å.$$

Body diagonal contact is $\frac{4.82}{\sqrt{3}} = 2.78Å$. The largest of these must be the length of the edge. Therefore, the correct one is 5.12 Å, from which

$$V = a^{3} = (5.12)^{3} = 134 \text{ Å}^{3}.$$

Volume of the ions is given by $4\left(\frac{4}{3}\pi\left(0.60^{3} + 1.81^{3}\right)\right) = 103Å^{3}.$

Therefore, the fraction void volume $= \frac{134 - 103}{134} = 1 - \frac{103}{134} = 0.23.$

Q. 5 : Determine the coordinate number for an atom in the body-centered cubic unit cell.

Fig.

Solution: Fig. 1 is an example of a body centered cubic cell. Now consider the atom in the center, it is surrounded by eight atoms. These are called the nearest nighbor atoms. In other words, the atom in the center of a body centered cubic unit cell is surrounded by eight nearest neighbor atoms.

By definition, the coordinate number of an atom in a crystal is the number of nearest-neighbor atoms. Therefore, a body centered cubic unit cell has a coordinate number of 8.

Q. 6: Potassium crystallizes with a body-centered cubic lattice and has a density of 0.856 g cm^{-3}. Calculate the length of the side of the unit cell a and the distance between (200), (110), and (222) planes.

Soution: (a) The relationship between the theoretical density of a unit cell and its dimensions is given by

$$d = \frac{CM}{AV} \quad (1)$$

Where d = density

c = the contents of the unit cell (that is the number of points contained within it).

M = the molecular weight of the substance that makes up the cell.

A = Avogadro's number.

V = volume, which is a^3 in this problem.

For a body-centered cubic unit cell, the contribution of the eight corners is 8 (1/8) = 1, and that of the body-centered atom is 1. Therefore, the contents of the unit cell, c = 1+1 = 2. Since the molecular weight of potassium is 39.098 g mol^{-1} and Avogadro's number,

$$A = 6.02 \times 10^{23}\ mol^{-1},$$

then equation (1) becomes

$$d = \frac{(2)(39.098\ g\ mol^{-1})}{(6.02 \times 10^{23}\ mol^{-1})\ a^3} \tag{2}$$

But d = 0.856 g cm^{-3}. Hence equation (2) can be written in the form

$$0.856\ g\ cm^{-3} = \frac{(2)(39.098g\ mol^{-1})}{(6.02 \times 10^{23}\ mol^{-1})\ a^3}.$$

Solving for a^3 yields

$$a^3 = \frac{(2)(39.098\ g\ mol^{-1})}{(6.02 \times 10^{23}\ mol^{-1})(0.856\ g\ cm^{-3})}$$

$$= \left[\frac{(78.196)}{5.15 \times 10^{23}}\right] cm^3$$

$$= 1.517 \times 10^{-22}\ cm^3$$

Therefore, a = 5.33 × 10^{-8} cm = 5.33 Å.

(b) The interplanar distance within a unit cell is related to its (unit cell) dimensions by

$$\frac{1}{d_{hkl}^2} = \frac{h^2}{a^2} + \frac{k^2}{b^2} + \frac{l^2}{c^2}.$$

But, in this problem a = b = c. Therefore

$$\frac{1}{d_{hkl}^2} = \frac{h^2}{a^2} + \frac{k^2}{b^2} + \frac{l^2}{c^2} = \frac{h^2 + k^2 + l^2}{a^2} \tag{3}$$

where d = interplanar distance between the Miller indices.
hkl = Miller indices.

Taking the reciprocal of equation (3) gives

$$= \frac{a^2}{h^2 + k^2 + l^2} \tag{4}$$

and taking the square root of equation (4) yields

$$\sqrt{d_{hkl}^2} = \frac{\sqrt{a^2}}{\sqrt{h^2 + k^2 + l^2}}$$

$$\therefore \quad d_{hkl} = \frac{a}{\sqrt{h^2 + k^2 + l^2}} \qquad (5)$$

For planes of (200) Miller indices, equation (5) gives

$$d_{200} = \frac{5.33\text{Å}}{\sqrt{2^2 + 0^2 + 0^2}}$$

$$= \frac{5.33\text{Å}}{\sqrt{4}} = \frac{5.33\text{Å}}{2} = 2.665\text{Å}.$$

For (110) planes,

$$d_{110} = \frac{5.33\text{Å}}{\sqrt{1^2 + 1^2 + 0^2}}$$

$$\therefore \quad d_{110} = \frac{5.33\text{Å}}{\sqrt{2}} = 3.77\text{Å}.$$

For (222) planes,

$$d_{222} = \frac{5.33\text{Å}}{\sqrt{2^2 + 2^2 + 2^2}}$$

$$d_{222} = \frac{5.33\text{Å}}{\sqrt{12}} = 1.540\text{Å}.$$

Q. 7 : List and explain the possible solid state defects that can occur in crystals.

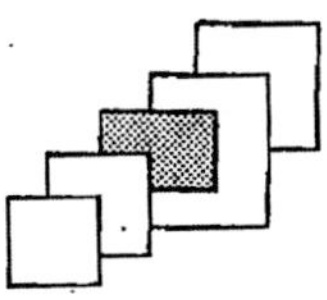

Fig. 1

Solution: The main type of solid state defects that can occur in crystals are lattice defects. There are three major types of lattice defects: lattice vacancies, lattice interstitials, and dislocations. Lattice vacancies are by far the most comon of the three. They arise when some of the lattice points are unocupied (some atoms are missing). This type of defect occurs to a small extent in all crystals, and properties such as diffusion and ionic conductivity can be explained by it.

Lattice interstitials arise if some atoms are pushed in and therefore occupy positions between lattice points. This occurs when small cations migrate into locations between normal planes of large anions.

The third type of lattice defect is dislocation. It consists of two types, edge dislocation and screw dislocation.

Edge dislocation comes about when an extra half-plane of atoms is inserted only partically in the upper half of a crystal as shown in Figure 1. The dislocation line is the shaded plane of the crystal.

As a result of this, the interatomic distances in the upper half crystal are compressed and those in the lower half-crystal are extended setting up tension. This effect is similar to moving a rug by pushing a crease down it.

Screw dislocation can be depicted when there is a fault line in a crystal and a resulting shift of the planes of the crystal relative to one another, as shown in Figure 2.

Fig. 2

Here the atoms to the left of the fault line are being pushed up through the distance of a unit cell. The unit cells of the crystal now form a continuous spiral around the end of the fault line, which is the screw axis, and breaks through to the surface where it takes the form of a spiral ramp. The dislocation line is along the axis of the unit cell.

Q. 8 : Consider the plane shown in the Fig. below which intersects the x–, y– and z– axes at a, b and c, the unit cell dimensions, respectively. What are the Miller indices for this plane ?

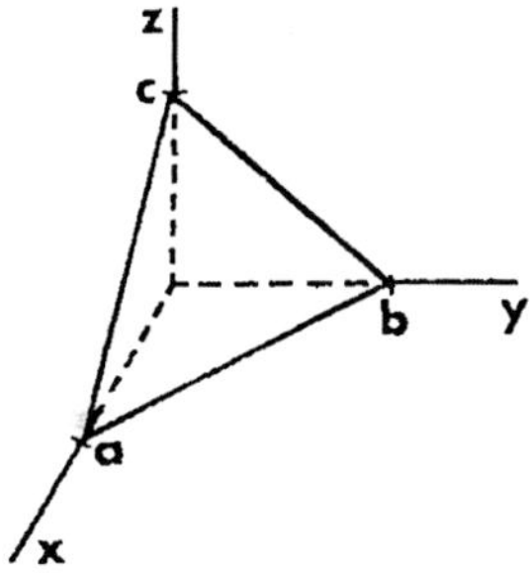

Fig.

Solution: Miller indices for a plane is prepared in the following manner :

(a) Make a three-column table with the three axes of the unit cell at the top of the columns.
(b) Enter in each column the intercept of the plane with that axis (this makes the first row).
(c) For the second row, invert all numbers.
(d) Then clear all fractions in the last (third) row.

Now if these steps are followed very carefully, a table as seen below would be obtained.

First row:

	1st column a	2nd column b	3rd column c	
	1	1	1	the intercepts.
Second row:	$\frac{1}{1} = 1$	$\frac{1}{1} = 1$	$\frac{1}{1} = 1$	inversion of the numbers, otherwise known as reciprocals.
Third row:	1	1	1	cleas fractions

Essentially, the last (third) row is the answer to our problem. (Make sure all fractions are cleared). The Miller indices (hkl) of this plane is thus (1 1 1).

Q. 9 : Find the spacing between the planes with indices 101 in NaCl if a = 5.6402 Å.

Solution: By definition, $\frac{1}{d_{hkl}^2} = \frac{h^2 + k^2 + l^2}{a^2}$ (1)

where d = spacing

hkl = Miller indices (101 respectively in this case). Substituting the respective values into (1), yields

$$\frac{1}{d_{101}^2} = \frac{1^2 + 0^2 + 1^2}{(5.6402\text{ Å})^2} = \frac{1}{31.81\text{ Å}^2}$$

$$d_{101}^2 = \frac{31.81\text{ Å}^2}{2}$$

$$\therefore \quad d_{101} = \sqrt{\frac{31.81\text{ Å}^2}{2}} = \frac{5.6402\text{ Å}}{1.4142} = 3.9883\text{ Å}.$$

Note that the general formula is

$$\frac{1}{a_{hkl}^2} = \frac{h^2}{a^2} + \frac{k^2}{b^2} + \frac{l^2}{c^2};$$

but in this problem, a = b = c.

Q. 10 : Determine the content for the body-centered cubic unit cell shown in the Fig. below.

Solution: There are eight corners in a body-centered cubic cell. The contribution of one corner is 1/8 of an atom. Therefore, the eight corners contribute :

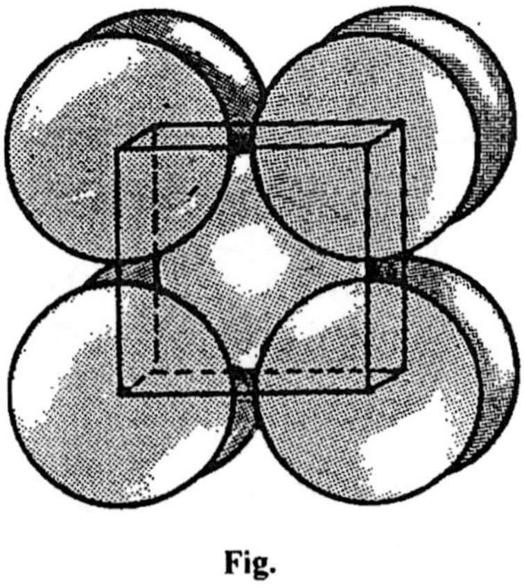

Fig.

$$\frac{1}{8}+\frac{1}{8}+\frac{1}{8}+\frac{1}{8}+\frac{1}{8}+\frac{1}{8}+\frac{1}{8}+\frac{1}{8}=8\left(\frac{1}{8}\right)=1.$$

We can also see that there is one atom in the center unshared by any corner. Consequently, there are

$$\left[8\left(\frac{1}{8}\right)+1\right] \text{atoms} = 2 \text{ atoms}$$

in a body-cenered cubic unit cell.

Q. 11 : Polonium is the only element known to crystallize in a primitive cubic unit cell under room conditions. If a = 3.36 Å, find the theoretical density.

Solution: Density, $d = \frac{ZM}{AV}$.

Z = content = 1

M = Molecular weight

A = Avogadro's number

V = Volume = a^3 in this case.

But $\quad V = (3.36\text{Å})^3 = (3.36\times10^{-10}\,\text{m})^3$

$$\therefore \quad d = \frac{(1)(209\ \text{gm/mole})(10^{-3}\,\text{kg/gm})}{(6.022\times10^{23}\,\text{mole}^{-1})(3.36\times10^{-10}\,\text{m})^3}$$

$$= 9.15\times10^{3}\,\text{kg/m}^3.$$

Note that a primitive unit cell has points only at the corners of a parallelepiped. So the total contribution from all corners = 1. This is why Z = content = 1, unlike others like edge-centered, body-centered or face centered cells that have points on the edge, body and face respectively (as suggested by their names.).

Q. 12 : Find the distance between two Po atoms that lie along a body diagonal if

a = 3.36 Å.

Solution: Po atoms form a cubic unit cell. The distance, L, between two atoms in a cubic unit cell is

$$L = \sqrt{a^2(x_2 - x_1)^2 + a^2(y_2 - y_1)^2 + a^2(z_2 - z_1)^2} \qquad (1)$$

where $(x_1y_1z_1)$ and $(x_2y_2z_2)$ are their respective coordinates. In this case $(x_1y_1z_1) = (000)$ and $(x_2y_2z_2) = (111)$. Substituting these values into equation (1) gives

$$L = \sqrt{(3.36)^2\left[(1-0)^2 + (1-0)^2 + (1-0)^2\right]} = 3.36\text{Å}\sqrt{3}$$

$$= (3.36\text{Å})(1.732) = 5.8197\text{Å} = 5.820\text{Å}.$$

Q. 13 : Determine the coordination number for an atom in the primitive cubic unit cell.

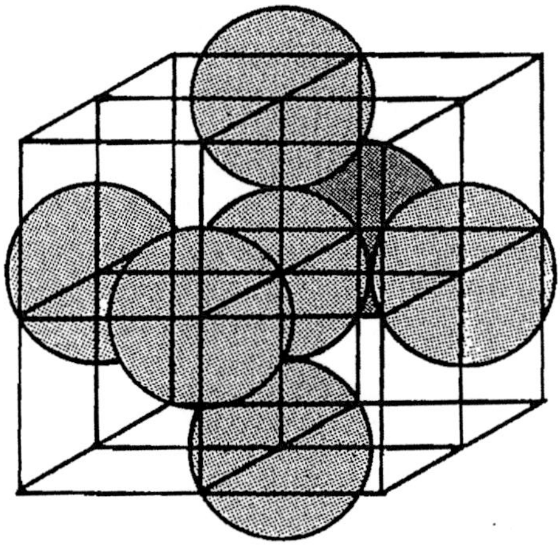

Fig.

Solution: If the primitive cubic unit cell were translated to generate several unit cells as shown in the fig. below around any given atom there would be six equally-spaced nearest-neighbor atoms at a distance a. Thus coordination number = 6. Any other atoms in the crystal lattice will be at a distance greater than a from the atom under consideration.

Q. 14 : Find the relationship between a and R for a primitive cubic unit cell and calculate R for Po if

$$a = 3.36 \text{ Å}.$$

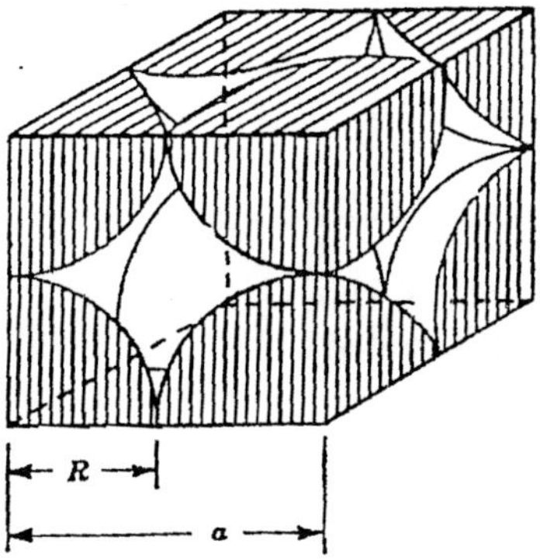

Fig. 1

Solution: By definition, R is the radius of an atom and a is the distance between the centers of two atoms which touch as shown in Fig. 1.

The relationship between a and R can be written as R = ½ a. Therefore a = 2R. Since

$$a = 3.36\text{Å},\ R = ½\,(3.36) = 1.68\ \text{Å}.$$

Q. 15 : Show that the void volume for structures of closest packed spheres is 25.9% in both ccp and hcp. For spheres of radius 0.100nm, calculate a_0 for ccp, and a_0, c_0 for hcp.

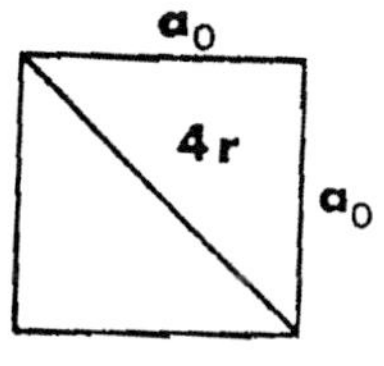

Fig.

Solution: (a) A cubic closest packed (ccp) unit cell has the same dimensions as the face centered cubic cell. For a face centered cubic, the length of the diagonal = 4r. Using the pythagorean theorem,

$$(4r)^2 = a_0^2 + a_0^2$$

$$16r^2 = 2a_0^2$$

or $$a_0 = \sqrt{\frac{16r^2}{2}} = \frac{4r}{\sqrt{2}} = \frac{2\sqrt{2}\sqrt{2}}{\sqrt{2}}\,r$$

$$= 2\sqrt{2}\, r.$$

The volume of this cell is therefore $\left(2\sqrt{2}r\right)^3 = a_0^3 = 8\left(\sqrt{2}\right)^3 r^3$

$$= (8)\,(2)\,\sqrt{2}\, r^3$$
$$= 16\sqrt{2}\, r^3.$$

Recall that the contents of a face centered unit cell add up to 4. That is, there are four spheres in the unit cell. Volume of a sphere $= \frac{4}{3}\pi r^3$. Therefore, the volume of 4 spheres is

$$4\left(\left(\frac{4}{3}\right)\pi r^3\right) = \left(\frac{16}{3}\right)\pi r^3.$$

The void volume percent is therefore

$$\frac{16\sqrt{2}\, r^3 - \left(\frac{16}{3}\right)\pi r^3}{16\sqrt{2}\, r^3} \times 100$$

$$= \left[\frac{16\sqrt{2}\, r^3}{16\sqrt{2}\, r^2} - \frac{\frac{16}{3}\pi r^3}{16\sqrt{2}\, r^3}\right] \times 100$$

$$= \left[1 - \frac{1\,\pi}{3\sqrt{2}}\right] \times 100$$

$$= 100 \frac{100\,\pi}{3\sqrt{2}} = 25.9\%$$

$$a_0 = 2\sqrt{2}\, r = 2\sqrt{2}(0.100 \text{ nm})$$
$$= 0.282 \text{ nm}.$$

(b) The hexagonal closest packed (hcp) unit cell has dimensions

$$a_0 = b_0 = 2r \text{ and } c_0 = \frac{4\sqrt{2}}{\sqrt{3}} r.$$

Volume of the cell = area of parallelogram × height

$$= (2r)^2 \sin 60° \times \frac{4\sqrt{2}}{\sqrt{3}} r$$

$$= 8\sqrt{2}\, r^3.$$

This unit cell is occupied by two spheres. Therefore, volume of both spheres $= 2\left(\left(\frac{4}{3}\right)\pi r^3\right) = \left(\frac{8}{3}\right)\pi r^3.$

The void volume percent is therefore

$$\left[\frac{8\sqrt{2}\, r^3 - \left(\frac{8}{3}\right)\pi r^3}{8\sqrt{2} r^3}\right] \times 100$$

$$= \left[\frac{8\sqrt{2}\, r^3}{8\sqrt{2}\, r^3} - \frac{\left(\frac{8}{3}\right)\pi r^3}{8\sqrt{2}\, r^3}\right] \times 100$$

$$= \left[1 - \frac{\pi}{3\sqrt{2}}\right] \times 100$$

$$= 100 - \frac{100\,\pi}{3\sqrt{2}} = 25.9\%.$$

$$a_0 = 2r = 2(0.100) = 0.200 \text{ nm}$$

$$b_0 = 2r = 0.200 \text{ nm}$$

$$c_0 = \frac{4\sqrt{2}}{\sqrt{3}} r = \frac{4\sqrt{2}}{\sqrt{3}}(0.100) = \frac{0.5656}{\sqrt{3}} = 0.327 \text{ nm}$$

Q. 16 : Magnesium oxide (M = 40.30) is cubic and has a density of 3.620 g. cm^{-3}. An X-ray diffraction diagram of MgO powder has lines at values of sin θ = 0.399, 0.461, 0.652, 0.764, 0.798, and 0.922. Index the pattern and determine the type of cubic structure. Calculate the wavelength of X-rays used. Assume that the number of MgO units per unit cell is the smallest consistent with the structure type.

Solution: This is a problem in which a test is conducted to determine the structure of magnesium oxide.

The test will be conducted by using the ITO method for indexing the X-ray powder pattern of the substance and inferring the dimensions of the cell.

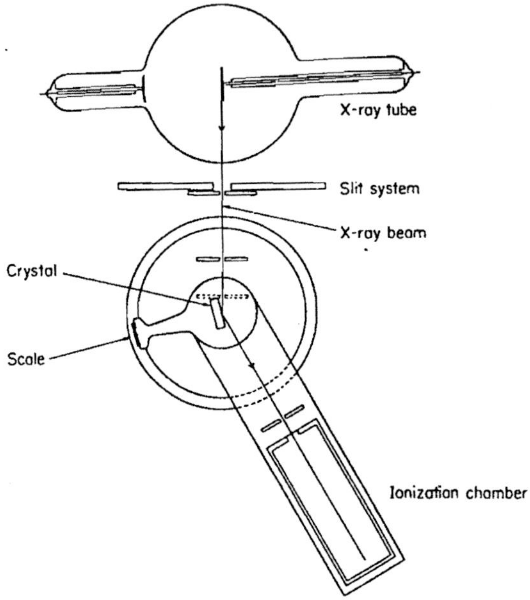

Bragg X-ray spectrometer.

Bragg developed a treatment of X-ray scattering by a crystal. He showed that the scatterings of X-rays could be represented as a "reflection" by successive planes of atoms in the crystal. All the waves "reflected" by a single crystal plane will be in phase and only under certain conditions will the waves "reflected" by different underlying planes be in phase with one another. The condition is that the difference in length of path between the waves scattered from successive planes much be an integral number of wavelengths, $n\lambda$. Thus the Bragg equation or "reflection" is

$$n\lambda = 2d \sin \theta$$

where λ = the wavelength

θ = angle of incident beam of X-rays, and n is an integer; i.e., n = 1,2,... The value of n determines its order. Hence, the "reflection" is first order if n = 1, second order if n = 2, etc. Suppose however, that θ is fixed. Hence sin θ is fixed. If the order is 1, then $\lambda = 2d \sin \theta$. In order to obtain the same value for λ when the order is 2, the spacing, d, must be ½ of what it was for first order.

To circumvent this problem the equation $n\lambda = 2d \sin \theta$ is rewritten as

$$\lambda = 2d_{hkl} \sin \theta. \quad (1)$$

Magnesium oxide could be based on one of the cubic lattices, simple, body-centered, or face-centered and by comparing the spacings calculated from the X-ray data with those expected for these lattices, the proper assignment of right cubic lattice can be made.

The distance between the planes (hkl) in a cubic lattice is given as

$$d_{hkl} = \frac{a_0}{\left(h^2 + k^2 + l^2\right)^{\frac{1}{2}}} \quad (2)$$

Combining equation (2) with equation (1) gives

$$\sin^2 \theta = \left(\frac{\lambda^2}{4a_0^2}\right)\left(h^2 + k^2 + l^2\right) \quad (3)$$

$\frac{\lambda^2}{4a_0^2}$ is a constant for the crystal; therefore

$$\frac{\sin^2 \theta}{\left(h^2 + k^2 + l^2\right)}$$

is also a constant. The observed value of sin θ can be indexed by assigning to it the value of (hkl) for the set of planes responsible for the "reflection".

Using the first 6 lines of each cubic structure, $\frac{\sin^2 \theta}{h^2 + k^2 + l^2}$ is computed for each and if its value is a constant for each of the (hkl), then it is that cell described by those Miller indices.

The first 6 lines of each cubic structure are :

(a) Simple cubic, (hkl) :	100	110	111	200	210	211
(b) Body centered, (hkl) :	110	200	211	220	310	222
(c) Face centered, (hkl) :	111	200	220	311	222	400

These values can be obtained for literatures.
Computing the

$$\frac{\sin^2\theta}{h^2+k^2+l^2}$$

value for the first two (hkl) values of each cubic structure gives

$$\text{simple cubic: (i)}\ \frac{\sin^2\theta}{h^2+k^2+l^2}=\frac{(0.399)^2}{1^2+1^2+0^2}=0.159$$

$$\text{(ii)}\ \frac{\sin^2\theta}{h^2+k^2+l^2}=\frac{(0.461)^2}{1^2+0^2+0^2}=0.106$$

$$\text{Body centered: (i)}\ \frac{\sin^2\theta}{h^2+k^2+l^2}=\frac{(0.399)^2}{1^2+1^2+0^2}=0.0795$$

$$\text{(ii)}\ \frac{\sin^2\theta}{h^2+k^2+l^2}=\frac{(0.461)^2}{2^2+0^2+0^2}=0.0531$$

$$\text{Face centered: (i)}\ \frac{\sin^2\theta}{h^2+k^2+l^2}=\frac{(0.399)^2}{1^2+1^2+1^2}=0.0530$$

$$\text{(ii)}\ \frac{\sin^2\theta}{h^2+k^2+l^2}=\frac{(0.461)^2}{2^2+0^2+0^2}=0.0530$$

The above values show that the $\frac{\sin^2\theta}{h^2+k^2+l^2}$ is a constant only for the face centered cubic structure. All the values are tabulated below.

		$\frac{\sin^2\theta}{h^2+k^2+l^2}$		
sin θ	$\sin^2\theta$	simple*	BCC*	FCC*
0.399	0.159	0.159	0.0795	0.0530
0.461	0.212	0.106	0.0531	0.0530
0.652	0.425	0.142	0.0708	0.0530
0.764	0.584	0.145	0.0732	0.0530
0.798	0.637	0.127	0.0637	0.0530
0.922	0.850	0.141	0.0707	0.0530

The constan value indicates that MgO is face centered cubic. Rearranging equation (3), the wavelength can be calculated. That is

$$\lambda = \sqrt{4a_0^2\left(\frac{\sin^2\theta}{h^2+k^2+l^2}\right)} \quad (4)$$

Before equation (4) can be used, a_0 needs to be known.

$$a_0 = V^{\frac{1}{3}} \text{ where } V = \text{volume.}$$

The volume, V can be otained from the relation

$$V = \frac{CM}{Ad} \quad (5)$$

where C = the contents of the unit cell.

M = Molecular weight

A = Avogadro's number

d = density.

For an fcc unit cell, C = 4. since all the parameters are known, equation (4) gives

$$V = \frac{(4)(40.30 \text{ g mol}^{-1})}{(6.02\times10^{23})\text{mol}^{-1}3.620 \text{ g cm}^{-3}}$$

$$= 73.94\times10^{-24}\text{cm}^3$$

$$= 0.07394 \text{ nm}^3$$

$$\therefore \quad a_0 = V^{\frac{1}{3}} = (0.07394 \text{ nm}^3)^{\frac{1}{3}} = 0.4197 \text{ nm.}$$

Substituting for a in equation (4) gives

$$\lambda = \sqrt{(4)(0.4197)^2(0.0530)}$$

$$= \sqrt{4(0.1761)\text{nm}^2 0.0530}$$

$$= \sqrt{0.0373 \text{ nm}^2}$$

or $\lambda = 0.1932$ nm

Q. 17 : Prepare a crystallographic projection for the body-centered cubic unit cell.

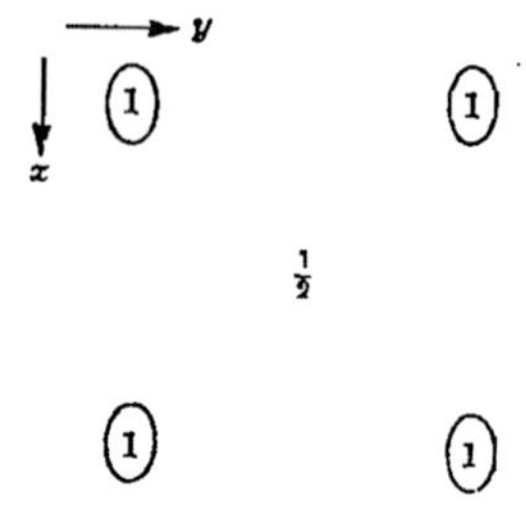

Solution: The eight corner atoms will be represented by four ① symbols at the corners of a square and the body-centered atom will be represented by a ½ in the center of the square, see Fig. below. The above is constructed by viewing a unit cell along one of the crystallographic axes. It shows the shape of a unit cell in two dimensions, with the three-dimensional information.

Q. 18 : If $\lambda = 1.5418$ Å for filtered Cu radiation, at what angle would the maximum reflection by the (200) plane of AgCl occur, assuming that

a =5.5491 Å ?

Solution: A beam of X-rays will be reflected if

$$\sin\theta = \frac{n\lambda}{2d} = \frac{\lambda}{2d_{hkl}} \qquad (1)$$

where θ = angle of reflection from the hkl plane

n = order of reflection (an integer).

λ = wavelength of the radiation

d_{hkl} = distance.

Looking at equation (1), n = 1 and $\lambda = 1.55418$ Å; but d_{hkl} is not given. This can be calculated using

$$\frac{1}{d^2_{hkl}} = \frac{h^2 + k^2 + l^2}{a^2}$$

where hkl = (200).

Therefore

$$\frac{1}{d^2_{200}} = \frac{h^2 + 0^2 + 0^2}{(5.5491\text{Å})^2} = 0.1299\text{Å}^{-2}$$

and

$$\frac{1}{d_{200}} = 0.3604\text{Å}^{-1}.$$

Thus $\quad d_{200} = \dfrac{1}{0.3604\text{Å}^{-1}} = 2.7746\text{Å}.$

Now that d_{hkl} is known, substitute its value into equation (1) and solve for θ. Hence

$$\sin\theta = \frac{1.5418\text{Å}}{(2)(2.7746\text{Å})} = \frac{1.5418}{5.5492} = 0.279$$

and $\quad \theta = \sin^{-1} 0.279$

$$\theta = 16.1°.$$

Q. 19 : A face-centered cubic structure gave the following powder pattern with X-rays of wavelength θ.1542 nm :

h k l	$\sin^2\theta$
1 1 1	0.0526
2 0 0	0.0701
2 2 0	0.1402
3 1 1	0.1928
2 2 2	0.2104

(a) Derive a general formula for $\sin^2\theta$ of a cubic substance in terms of wavelength, the unit cell dimension, and the Miller indices.

(b) Calculate the interplanar spacing for each of the preceding lines.

(c) From each of these lines, determine a value for the length of the unit cell edge.

Solution: (a) The interplanar distance (or spacing) is related to the dimensions of a unit cell by the formula

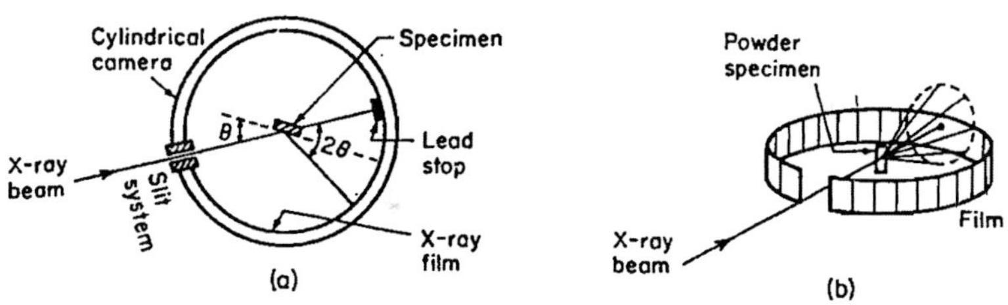

X-ray diffraction by the powder method.

$$\frac{1}{d_{hkl}^2}=\frac{h^2}{a^2}+\frac{k^2}{b^2}+\frac{l^2}{c^2} \qquad (1)$$

where d_{hkl} = interplanar distance

a, b and c = dimensions of the unit cell

h, k and l = Miller indices.

In this particular cell under consideration, a = b = c. Therefore, equation (1) changes to

$$\frac{1}{d_{hkl}^2}=\frac{h^2}{a^2}+\frac{k^2}{a^2}+\frac{l^2}{a^2}=\frac{h^2+k^2+l^2}{a^2} \qquad (2)$$

Taking the square root and reciprocal of equation (2) gives

$$\sqrt{d_{hkl}^2}=\sqrt{\frac{a^2}{h^2+k^2+l^2}}$$

$$\therefore \qquad d_{hkl}=\frac{a}{\sqrt{h^2+k^2+l^2}} \qquad (3)$$

The Bragg equation is given by

$$\lambda = 2\,(d/n)\sin\theta \qquad (4)$$

where n = an integer known as the order of reflection.

λ = wavelength of the readiation.

In equation (4), the d/n = d_{hkl}; which means that higher order reflections can be treated as first order reflections from planes with spacing d/n.

Thus, $\lambda = 2d_{hkl}\sin\theta$. (5)

But
$$d_{hkl}=\frac{a}{\sqrt{h^2+k^2+l^2}}$$

Therefore, equation (5) becomes

$$\lambda = 2\frac{a}{\sqrt{h^2+k^2+l^2}}\sin\theta \qquad (6)$$

Squaring both sides of equation (6) gives

$$\lambda^2=\left[2\frac{a}{\sqrt{h^2+k^2+l^2}}\sin\theta\right]^2$$

or
$$\lambda^2 = 4\frac{a}{\sqrt{h^2+k^2+l^2}}\sin\theta. \qquad (7)$$

Rarranging equation (7) to solve for $\sin^2\theta$ gives

$$\sin^2\theta = \frac{\lambda^2}{4a^2}\left(h^2 + k^2 + l^2\right). \qquad (8)$$

(b) Reacall that the $\frac{h^2 + k^2 + l^2}{a^2}$ term in equation (8) is $\frac{1}{d_{hkl}^2}$.

Therefore, rewriting it in terms of $\frac{1}{d_{hkl}^2}$ gives

$$\sin^2\theta = \frac{\lambda^2}{4d_{hkl}^2} \qquad (9)$$

Solving equation (9) for d_{hkl}^2 gives

$$d_{hkl}^2 = \frac{\lambda^2}{4\sin^2\theta} \qquad (10)$$

Taking the square root of both sides, equation (10) becomes

$$d_{hkl} = \frac{\lambda}{2\sin\theta} = \frac{0.1542 \text{ nm}}{2\sin\theta} \qquad (11)$$

Putting the sin θ value of the first row of the table in equation (11) gives

$$d_{hkl} = \frac{0.1542 \text{ nm}}{2\sqrt{\sin^2\theta}} = \frac{0.1542 \text{ nm}}{(2)\sqrt{0.0525}}$$

$$= \frac{0.1542 \text{ nm}}{(2)(0.2293)} = 0.3362 \text{ nm}.$$

The others are tabulated below :

hkl	11	200	220	311	222
$\sin\theta$	0.2293	0.2647	0.3744	0.4391	0.4587
d_{hkl}, nm	0.3362	0.2913	0.2059	0.1756	0.1681

(c) Rearranging equation (3) to solve for a gives

$$a = d_{hkl}\left(\sqrt{h^2 + k^2 + l^2}\right) \qquad (12)$$

Using the different d_{hkl} values and their corresponding hkl values in equation (12), a can be determined.That is

$$a = 0.3362\left(\sqrt{1^2+1^2+1^2}\right) = 0.582 \text{ nm}$$

$$a = 0.2913\left(\sqrt{2^2+0^2+0^2}\right) = 0.582 \text{ nm}$$

$$a = 0.2059\left(\sqrt{2^2+2^2+0^2}\right) = 0.582 \text{ nm}$$

$$a = 0.1756\left(\sqrt{3^2+1^2+1^2}\right) = 0.582 \text{ nm}$$

$$a = 0.1681\left(\sqrt{2^2+2^2+2^2}\right) = 0.582 \text{ nm}$$

Therefore, the required value is 0.582 nm.

12
Nuclear Chemistry

Q. 1 : 2.000 picogram (pg) of ^{33}p decays by $^{0}_{-1}\beta$ emission to 0.250 pg in 75.9 days. Find the half-life of ^{33}p.

Solution: The half-life is defined as the time it takes for ½ of the amount of a certain compound present to decompose. For example, if a substance has a half-life of 1 day, after one day there will only be ½ of the original amount left. When given the original and final amount of a substance, after a given time has elapsed, the number of half-lives that have passed can be found. This is done by dividing the original amount by the final amount and determining how many factors of 2 are present in the quotient. The half-life is then found by dividing the time elapsed by the number of twos. Mathematically, this is done as follows :

Solving for the half-life of ^{33}p :

$$\frac{\text{original amount}}{\text{final amount}} = \frac{2.000 \text{ pg}}{0.250 \text{ pg}} = 8$$

Obtaining the number of 2's in the quotient yields

$8 = 2 \times 2 \times 2$; Therefore 3 half-lives have passed.

One is given that these half-lives elapse in 75.9 days. Consequently, the half-life of ^{33}p is

$$\text{half} - \text{life} = \frac{75.9 \text{ days}}{3} = 25.3 \text{ days.}$$

Q. 2 : One curie is defined as the amount of material, which is equaivalent to one gram of radium, that gives 3.7×10^{10} nuclear disintegrations per second. The half-life of ^{210}At is 8.3 hr. How many grams of astatine (At) would equal one curie ?

Solution: The half life is defined as the time it takes for one half of the amount of a radioactive substance to disintegrate. To find the number of grams of ^{210}At that would be equal to 1 curie, first find the number of disintegrations of ^{210}At per second. Then solve for the number of moles necessary and from that solve for the number of grams.

Solving :

(1) determine the half-life in seconds. This unit is used because curies are measured in disintegration per second.

$$t_{1/2} = 8.3 \text{ hr} \times 60 \text{ min/hr} \times 60 \text{ sec/min} = 29880 \text{ sec}$$

(2) Solving for the number of disintegrations of ^{210}At in 1 curie.

Let x = the total number of nuclei of ^{210}At in 1 curie, x/2 the number of nuclei left after 1 half-life, 29880 sec. The following ratio can be set up :

$$\frac{(^{210}\text{At}) \text{ number of nuclei after 1 half - life}}{\text{half - life}}$$
$$= 3.7 \times 10^{10} \text{ dis/sec.}$$

One can solve this ratio for x, the original number of nuclei present.

$$\frac{x/2}{29880 \text{ sec}} = 3.7 \times 10^{10} \text{dis/sec}$$

$$\frac{x}{2} = 29880 \text{ sec} \times 3.7 \times 10^{10} \text{dis/sec}$$

$$x = 2 \times 29880 \text{ sec} \times 3.7 \times 10^{10} \text{ dis/sec} = 2.21 \times 10^{15} \text{ dis.}$$

Thus, 2.21×10^{15} dis of ^{210}At are equal to 1 curie.

(3) Solving for the number of moles. There are 6.02×10^{23} nuclei per mole. Therefore, the number of moles present is equal to the number of disintegrations divided by 6.02×10^{23}.

$$\text{no. of moles} = \frac{2.21 \times 10^{15} \text{nuclei}}{6.02 \times 10^{23} \text{nuclei / mole}} = 3.67 \times 10^{-9} \text{moles}$$

(4) On solves for the number of grams necessary by multiplying the number of moles by the molecular weigth of ^{210}At (MW = 210).

no. of grams = 210 g/mole × 3.67×10^{-9} moles

$= 7.71 \times 10^{-7}$ g.

This would be the number of grams necessary at $t = t_{1/2}$.

Q. 3 : The half-life, $t_{1/2}$ of $^{220}_{86}Rn$ is given as 54.5 sec. Calculate the mass of $^{220}_{86}Rn$ which is equivalent to 1 millicurie.

Solution: The rate of emission from a sample of a particular radioactive substance is proportional to the number of nuclei remaining, consequently the rate is continuously decreasing. Expressing this mathematically gives

$$\frac{-dN}{dt} = \lambda N \qquad (1)$$

which reads :

number decaying per sec = a proportionality constant λ, characteristic of the isotope times the number of nuclei remaining, N.

If at some particular time there are N_0 nuclei in the sample, then an expression can be found for N, the number remaining at any later time t, by integrating equation (1). That is,

$$\int -\frac{dN}{N} = \int \lambda \, dt$$

$-\ln N = \lambda t +$ constant of integration. To evaluate the constant, let N_0 be the number of nuclei present at some arbitrary zero time,

When $t = 0$

$-\ln N_0 =$ constant of integration

Therefore $-\ln N_0 = \lambda t - \ln N_0$

or $\ln N_0 - \ln N = \lambda t$

$$in(N_0 / N) = \lambda t$$

Therefore $$\frac{N_0}{N} = e^{\lambda t}$$

or $$\frac{N}{N_0} = e^{-\lambda t} \qquad (2)$$

and $\ln (N/N_0) = \ln e^{-\lambda t}$

$$= -\lambda t \qquad (3)$$

A relationship between the decay constant λ and the half-life $t_{1/2}$ can be found from equation (3). At time $t_{1/2}$, the number of nuclei remaining is exactly one half of the initial number, so that the ratio

N/N_0 is numerically equal to one half

$$t_{½} = [-\text{in } ½]/\lambda$$
$$= [\text{in}(1/½)]/\lambda$$
$$= [\text{in } 2]/\lambda$$
$$= 0.693/\lambda$$

Therefore $t_{½} = \dfrac{0.693}{\lambda}$ (4)

from which $\lambda = \dfrac{0.693}{t_{½}}$

$$= \frac{0.693}{54.5 \text{ sec}}$$
$$= 1.27 \times 10^{-2} S^{-1}.$$

Using equation (1) and the definition of a curie given as

$$\frac{-dN}{dt} = 3.7 \times 10^7 \text{ sec}^{-1},$$

N can be calculated as follows :

$$N = \frac{-(dN/dt)}{\lambda}$$
$$= \frac{3.7 \times 10^7 \text{ sec}^{-1}}{1.27 \times 10^{-2} \text{ sec}^{-1}}$$
$$= 2.91 \times 10^9$$

Converting this number to moles and mass gives :

$$\left(\frac{2.91 \times 10^9}{6.02 \times 10^{23} \text{mol}^{-1}}\right)(220 \text{ g mol}^{-1})$$
$$= 106 \times 10^{-14} \text{ g} = 106 \times 10^{-17} \text{kg}$$

or $= 106 \times 10^{-15} \text{kg}$

Q. 4 : A slow neutron flux of 10^{12} neutrons per cm^2 bombards a sample of cobalt for a period of six months. Given that the cross section of ^{59}Co is 21.7 barns and ^{60}Co has a half-life of 5.26 years, calculate the net activity of the cobalt sample.

Solution: The average rate of radioactive decay is proportional to the number of radioactive atoms, N, present and is given by

$$-\frac{dN}{dt} = \lambda N \tag{1}$$

where λ = the decay constant analogous to the first-order rate constant. The rate of formation of the nuclides is given by the product of the cross section σ, the neutron flux density φ and the concentration N_0. That is, rate of formation = $\sigma\varphi N_0$.

Therefore the net rate of formation is

$$\frac{dN}{dt} = \sigma\varphi N_0 - \lambda N \tag{2}$$

$$= R - \lambda N$$

where $R = \sigma\varphi N_0$. (3)

or

$$\frac{dN}{R - \lambda N} = dt \tag{4}$$

Integrating equation (4) between t = 0, N = 0 and t = t, N = N gives

$$\int_{N=0}^{N=N} \frac{dN}{R - \lambda N} = \int_{t=0}^{t=t} dt$$

$$-\frac{1}{\lambda} \ln \frac{R - \lambda N}{R} = t$$

Solving for N gives

$$-\lambda t = \ln\left(\frac{R - \lambda N}{R}\right)$$

$$= \ln\left(1 - \frac{\lambda N}{R}\right)$$

$$e^{-\lambda t} = e^{\ln(1 - \lambda N/R)}$$

$$= 1 - \frac{\lambda N}{R}$$

$$\frac{\lambda N}{R} = 1 - e^{-\lambda t}$$

or

$$N = \frac{R\left(1 - e^{-\lambda t}\right)}{\lambda} \tag{5}$$

Recall that $R = \sigma\varphi N_0$. Therefore, substituting for R in equation

(5) gives

$$N = \frac{\sigma\varphi N_0(1-e^{-\lambda t})}{\lambda}$$

$$= \frac{\sigma\varphi N_0}{\lambda}(1-e^{-\lambda t}) \qquad (6)$$

$$N_0 = \left(\frac{8.9 \text{ g } ^{60}\text{Co}}{\text{cm}^3}\right)\left(\frac{1 \text{ mol}}{59 \text{ g}}\right)\left(\frac{6.02 \times 10^{23} \text{ molecules}}{1 \text{ mole}}\right)$$

$$= 9.08 \times 10^{22} \text{ molecules cm}^{-3}.$$

But $\quad t_{½} = \dfrac{0.693}{\lambda}$

where $\quad t_{½}$ = half-life

$\therefore \quad \lambda = \dfrac{0.693}{t_{½}}$

$$= \frac{0.693}{(5.26\text{years})\left(\frac{365 \text{ days}}{1 \text{ year}}\right)\left(\frac{24 \text{ hrs}}{1 \text{ day}}\right)\left(\frac{3600 \text{ sec}}{1 \text{ hr}}\right)}$$

$$= 4.2\times10^{-9} \text{ sec}^{-1}$$

Using equation (6) yields

$$N = (2.17 \text{ barns})\left(\frac{10^{-24} \text{ cm}^2}{1 \text{ barn}}\right)\left(\frac{10^{12}}{\text{cm}^2 - \text{sec}}\right)$$

$$(9.08 \times 10^2 \text{ cm}^{-3})\left(1 - e^{-0.693t/5.6}\right)$$

$$= 8.82 \times 10^{19} \frac{\text{atoms}}{\text{cm}^3}.$$

From equation (1),

$$-\frac{dN}{dt} = (4.2 \times 10^{-9} \text{sec}^{-1})\left(8.82 \times 10^{19} \frac{\text{atoms}}{\text{cm}^3}\right)\left(\frac{1 \text{ cm}^3}{9 \text{ g}}\right)$$

$$= 4.11 \times 10^{10} \frac{\text{atoms}}{\text{sec}}.$$

Therefore, the number of radioactive material or the curies in the ^{60}Co sample is obtained by using the definition of the curie, which is 3.7×10^{10} disintegrations per second.

That is, $$\frac{4.11 \times 10^{10}}{3.7 \times 10^{10}}$$

$$= 1.10 \frac{\text{curies}}{\text{g}} \times \text{number of grams.}$$

Note that the value 1.10 curies/g is the specific activity.

Q. 5 : N^{13} decays by β^+ emission. The maximum kinetic energy of the β^+ is 1.19 MeV. What is the nuclidic mass of N^{13}?

Solution: In the emission of a β^+ particle, the daughter nuclide has a Z value one unit less than the parent with no change in A (mass). Thus,

$$_ZP^A \rightarrow _{Z-1}D^A + \beta^+,$$

where P and D are the parent and daughter nuclides, respectively. The emitted positron (β^+) is unstable and is usually consumed after being slowed down by collisions. This is shown in the following reaction

$$\beta^+ + \beta^- \rightarrow 2\gamma.$$

The gamma (γ) radiation is a form of light. It has no mass and no charge.

In this reaction, 2 photons of light are produced.

The reaction in this problem is

$$_7N^{13} \rightarrow {}_6C^{13} + \beta^+$$

In this type of process, a simple difference of whole atom masses is not desired. Whole atom masses can be used for mass difference calculations in all nuclear reactions, except in β^+ processes where there is a resulting annihilation of two electron masses (one β^+ and one β^-). Thus,

$$\begin{aligned}\text{mass difference} &= (M_n \text{ for } {}_7N^{13}) - (M_n \text{ for } {}_6C^{13}) - M_e \\ &= [(M \text{ for } {}_7N^{13}) - 7M_e] - [(M \text{ for } {}_6C^{13}) - 6M_e] - M_e \\ &= (M \text{ for } {}_7N^{13}) - (M \text{ for } {}_6C^{13}) - 2M_e \\ &= (M \text{ for } {}_7N^{13}) - 13.00335 - 2(0.00055) \\ &= (M \text{ for } {}_7N^{13}) - 13.00.445\end{aligned}$$

Here, M_n is the nuclear mass (mass of neutrons and protons),

M is the atomic mass, and M_e is the mass of an electron. This expression is equal to the mass equivalent of the maximum kinetic energy of the β^+.

$$\frac{1.19 \text{ MeV}}{931.5 \text{ MeV/amu}} = 0.00128 \text{ amu}$$

Then, 0.00128 amu = (M for $_7N^{13}$) – 13.00445 or,
M for $_7N^{13}$ = 13.00445 + 0.00128 = 13.00573 amu.

Q. 6 : Calculate Δm, the difference in mass between the final and initial nuclei in grams per mole for the emission of a γ ray, $^{19}_{8}O^*$ (excited state)$\rightarrow^{19}_{8}$ O(ground state) $+\gamma(\Delta E = 1.06 \times 10^8$ kcal / mole).

Solution: Δm can be solved for by using Einstein's Law, which relates Δm and ΔE.

This law is given by

$$\Delta E = \Delta mc^2,$$

where ΔE = the energy, Δm is the change in mass, and c is the speed of light, 3×10^{10} cm/sec. When using Einstein's Law, the ΔE, in kcal/mole, must be first converted to ergs/mole. There are 4.18×10^{10} ergs in one kcal, therefore one converts kcal to ergs by multiplying the number of kcal by 4.18×10^{10} ergs/kcal.

$$\text{no. of ergs} = 4.18 \times 10^{10} \text{ ergs/kcal} \times 1.06 \times 10^8 \text{ kcal/mole}$$
$$= 4.43 \times 10^{18} \text{ ergs/mole.}$$

Solving for Δm,

$$\Delta m = \frac{\Delta E}{c^2}, \text{ where } \Delta E = 4.43 \times 10^{18} \text{ ergs / mole}$$

$$c = 3 \times 10^{10} \text{ cm / sec}$$

$$\text{ergs} = \frac{\text{g cm}^2}{\text{sec}^2}$$

$$\therefore \quad \Delta m = \frac{4.43 \times 10^{18}}{(3 \times 10^{10})^2} \frac{\text{g}}{\text{mole}}$$

$$\Delta m = 4.92 \times 10^{-3} \text{ g / mole}$$

When a γ particle is emitted, one knows that the final state weighs less than the initial state, therefore,

$$\Delta m = -4.92 \times 10^{-3} \text{ g / mole.}$$

Q. 7 : Calculate the weight, W, in grams of 1.00 millicurie of ^{14}C from its half life of 5720 years.

Solution: The decay of a radioactive substance follows the exponential law

$$N = N_0 e^{-\lambda t}, \qquad (1)$$

where N = the number of unchanged atoms at time t, N_0 is the number present at t = 0, and λ is the characteristic decay constant for the substance. The characteristic rate of radioactive decay may be given in terms of the half-life $t_{1/2}$, which is the time required for an initial number of atoms N, to be reduced in half. Thus, at the time $t = t_{1/2}$, $N = N_0/2$. Equation (1) thus becomes

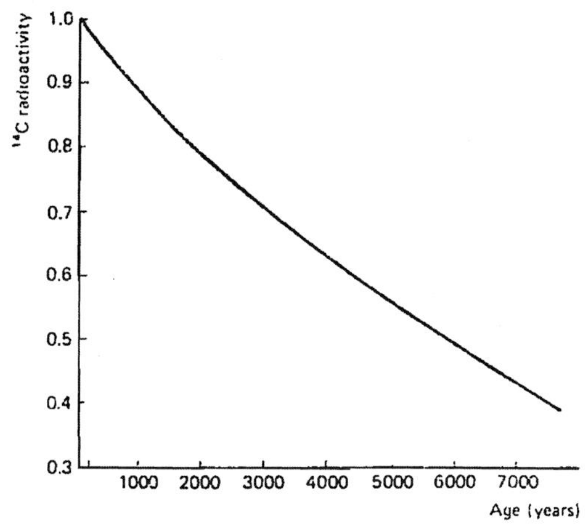

$$\frac{N_0}{2} = N_0 e^{-\lambda t_{1/2}}$$

or $$\frac{1}{2} = e^{-\lambda t_{1/2}}$$

and $$\ln\frac{1}{2} = -\lambda t_{1/2}.$$

from which $$t_{1/2} = \frac{\ln 2}{\lambda}$$

$$= \frac{0.693}{\lambda} \qquad (2)$$

If the deacy rate, $-\frac{dN}{dt}$, is set proportional to the number of atoms present: then

$$-\frac{dN}{dt} = \lambda N$$

$$= \frac{\lambda \text{ W in g}}{14 \text{ g mole}^{-1}} \times 6.02 \times 10^{23} \text{mole}^{-1} \qquad (3)$$

Solving for λ in equation (2) yields

$$\lambda = \frac{0.693}{t_{1/2}}$$

$$= \frac{0.693}{(5720 \text{ years})\left(\frac{365 \text{ days}}{1 \text{ year}}\right)\left(\frac{24 \text{ hrs}}{1 \text{ day}}\right)\left(\frac{3600 \text{ sec}}{1 \text{ hr}}\right)}$$

$$= 3.83 \times 10^{-12} \text{ sec}^{-1}.$$

But $\frac{-dN}{dt} = 1.00$ millicurie

$= 3.700 \times 10^{7}$ disintegration per second.

Therefore, from equation (3),

$$W = \frac{-dN/dt\,(14)}{(3.83 \times 10^{7} \text{ sec}^{-1})(6.02 \times 10^{23})}$$

$$= 2.24 \times 10^{-4} \text{g}.$$

Q. 8 : How many alpha particles per second would be emitted from 4×10^{-12} g of ^{210}Po ($t_{1/2}$ = 138 days) ?

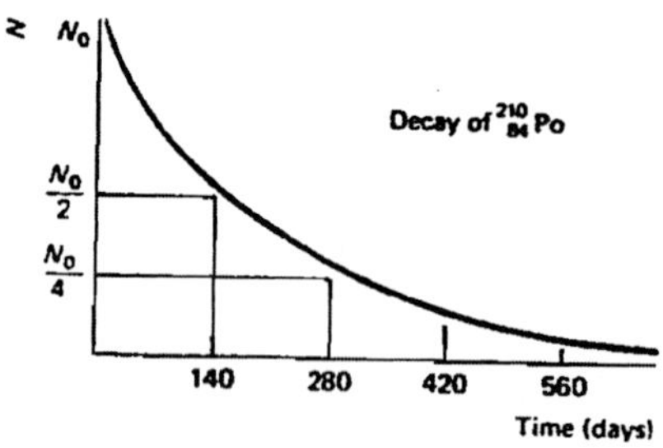

Solution: When given the mass and the half-life of a substance, one can find the number of disintegrations per second (dN/dt) by using the following equation :

$$\frac{dN}{dt} = -\frac{\ln 2}{t_{1/2}} \times N,$$

where $t_{1/2}$ is the half-life in second, N is the original number of nuclei.

(1) Solving for the half-life in seconds :

$$t_{1/2} = 138 \text{ days x } 24 \text{ hr/day x } 60 \text{ min/hr x } 60 \text{ sec/min}$$
$$= 1.192 \times 10^7 \text{ sec.}$$

(2) Determining the number of nuclei present. One is given that there was originally 4×10^{-12}g of ^{210}Po present. The number of nuclei present is found by multiplying the number of moles by Avogrado's number (6.02×10^{23}), the number of nuclei in one mole. The number of moles is found by dividing 4×10^{-12}g by the molecular weight of ^{210}Po. (MW = 210).

$$\text{no. of moles} = \frac{4 \times 10^{-12}\text{g}}{210 \text{ g / mole}} = 1.905 \times 10^{-14} \text{ moles}$$

$$\text{no. of nuclei} = 1.905 \times 10^{-14} \text{ moles} \times 6.022 \times 10^{23} \text{ nuclei / mole}$$
$$= 1.147 \times 10^{10} \text{ nuclei}$$

(3) Solving for dN/dt,

$$\frac{dN}{dt} = -\frac{\ln 2}{t_{1/2}} \times N \qquad N = 1.147 \times 10^{10} \text{ nuclei}$$
$$t_{1/2} = 1.192 \times 10^7 \text{ sec}$$
$$\ln 2 = 0.693$$

$$\frac{dN}{dt} = -\frac{0.693}{1.192 \times 10^7 \text{ sec}} \times 1.147 \times 10^{10} \text{ nuclei}$$
$$= -666.94 \text{ nuclei / sec}$$

Because dN/dt is negative, it means that the change in the number of nuclei per unit time is caused by disintegrations of the nuclei. This is the disintegration rate at t = 0.

Q. 9 : A lead foil 1 cm^2 in area and 1 mm thick is bombarded by a flux of thermal neutrons equal to 10^8 sec^{-1} cm^{-2}. Given that the cross section, σ of lead is 0.18 barn and its density = 11.3×10^3 kg m^{-3}, calculate the number of radioactive nuclei formed per second.

Solution: For a beam of particles striking a given target, the number of nuclei formed is given by the equation

$$I_0 - I = I_0\left(1 - e^{-n\sigma_f x}\right) \qquad (1)$$

where I = the intensity of the beam after passing through a target thickness x, I_0 is the incident intensity and $I_0 - I$ is the number of reactions per unit time. n is the number of nuclei present per m^3, and since the atomic weight of lead is given, n is equal to (weight of lead) × (Avogadro's number) ÷ the atomic weight. Therefore

$$n = \frac{\left(11.3 \times 10^3 \text{ kg m}^{-3}\right)\left(6.02 \times 10^{23} \text{ nuclei mol}^{-1}\right)}{\left(207 \text{ g mol}^{-1}\right)\left(\dfrac{10^{-3} \text{ kg}}{1 \text{ g}}\right)}$$

$$= 3.29 \times 10^{28} \text{ nuclei / m}^3$$

$$x = 1 \text{ mm}\left(\frac{10^{-3}\text{m}}{1 \text{ mm}}\right)$$

$$= 1 \times 10^{-3}\text{m}.$$

The intensity of the incident beam is

$$I_0 = (\text{flux}) \times (\text{area})$$

$$= \left(10^8 \text{ sec}^{-1}\text{cm}^{-2}\right) \times \left(1 \text{ cm}^2\right)$$

$$= 10^8 \text{ sec}^{-1}.$$

Therefore, from equation (1)

$$I_0 - I = 10^8[1 - \exp(-3.29 \times 10^{28}\text{m}^{-3})\left(0.18 \times 10^{-24}\text{cm}^2\right)\left(\frac{1 \text{ m}}{100 \text{ cm}}\right)^2\left(1 \times 10^{-3}\text{m}\right)]$$

$$= 10^8\left[1 - \exp\left(-5.9 \times 10^{-4}\right)\right]$$

$$= 5.9 \times 10^4 \text{ lead nuclei formed per second.}$$

Q. 10 : A sample of bristle cone pine wood of age 7,000 ± 100 years, known by counting growth rings, possesses an activity of 6.6 d (disintegrations)/minute-g of carbon. Calculate the exact age of the wood sample from radiochemical evidence.

Solution: Radioactive decay is related to the rate of disintegration and the half-life by the following equation :

$$q_t = q_0(0.5)^{t/t_{1/2}}$$

where q_t is the rate of disintegration of the substance containing carbon, q_0 is the rate of disintegration of carbon (14 d/min-g), t is the time elapsed during the reaction, and $t_{½}$ is the half-life of the element concerned. For carbon, $t_{½}$ = 5730 yrs. Here one needs to solve for t.

$$q_t = q_0(0.5)^{t/t_{½}} \qquad q_t = 6.6\ d/min-g$$

$$q_0 = 14\ d/min-g$$

$$t = ?$$

$$t_{½} = 5730\ yrs$$

$$6.6\ d/min-g = 14\ d/min-g \times (0.5)^{t/5730}$$

$$\frac{6.6\ d/min-g}{14\ d/min-g} = 0.5^{t/5730}$$

$$.471 = 0.5^{t/5730}$$

$$\log .471 = t/5730\ yrs \times \log 0.5$$

$$-.327 = t/5730 \times (-.301)$$

$$\frac{-.327}{-.301} = \frac{t}{5730\ yrs}$$

$$t = \frac{.327 \times 5730\ yrs}{.301} = 6225.0\ yrs.$$

Therefore, the tree is 6,225 years old.

Q. 11 : The radioactive decay constant for radium is 1.36×10^{-11}. How many disintegrations per second occur in 100 g of radium? The molecular weight of radium is 226 g/mole.

Solution: The number of disintegration per second of a given amount of a particular element can be determined by using the following equation.

$$D = \lambda N$$

where D = the number of disintegration per second, λ is the decay constant and N is the number of atoms present.

In this problem λ is given and N must be determined before solving for D. Given that 100 g of radium is present, the number of moles present is determined by dividing 100 g by the molecular weight of radium. Therefore,

$$\text{no. of moles} = \frac{100\ g}{226\ g/mole} = 0.442\ \text{moles.}$$

There are 6.02×10^{23} particles per mole; Thus, calculate the number of atoms (N) in 0.442 moles an follows :

$$N = (0.442 \text{ moles}) (6.02 \times 10^{23} \text{ atoms/mole})$$
$$= 2.66 \times 10^{23} \text{ atoms}$$

Consequently, solving for D yields

$$D = \lambda N = (1.36 \times 10^{-11}) (2.66 \times 10^{23}) = 3.62 \times 10^{12} \text{ dis/sec.}$$

Q. 12 : Calculate the maximum kinetic energy of the β^- emitted in the radioactive decay of He^6.

Solution: To solve this problem, one must first know that β^- is an electron of negative unit charge and zero mass. thus, the equation for this particular decay process is

$$^6_2He \rightarrow {}^6_3Li + \beta^-. \text{ or } {}_2He^6 \rightarrow {}_3Li^6 + {}_{-1}e^0.$$

The kinetic energy of the β^- particle comes from the difference in mass (or energy, $E = mc^2$), between products and reactants. To compute the mass change during this process, only the whole atomic mass of He^6 and Li^6 need be considered. Once the difference in mass is known, a conversion factor is used (931. MeV = 1 amu). Thus

Mass of He^6	=	6.01890 aum
Mass of Li^6	=	6.01513 aum
loss in mass	=	0.00377 aum

Energy equivalent=(931 Me V/1 amu) (0.00377 amu)=3.51 MeV.

Therefore, the Maximum kinetic energy of the β^- particle is 3.51 MeV..

Q. 13 : The first step in the radioactive decay of $^{238}_{92}U$ is $^{238}_{92}U = {}^{234}_{90}Th + {}^4_2He$. Calculate the energy released in this reaction. The exact masses of $^{238}_{92}U$, $^{234}_{90}Th$, and ^4_2He are 238.0508, 234.0437 and 4.0026 amu, respectively. (1.0073 amu = 1.673×10^{-24} g.)

Solution: The energy released in this process can be determined from the change in mass that occurs. Energy and mass are related by the following equation,

$$\Delta E = \Delta mc^2$$

where, ΔE is the change in energy, Δm is the change in mass and c is the speed of light (3.0×10^{10} cm/sec). Δm is found by subtracting the mass of $^{238}_{92}U$ from the sum of the masses of $^{234}_{90}Th$ and ^4_2He.

$$\Delta m = (234.0437 \text{ amu} + 4.0026 \text{ amu}) - 238.0508 \text{ amu}$$
$$= -.0045 \text{ amu.}$$

(Note that amu is called atomic mass unit, and is defined such that the mass of the particular carbon atom having six neutrons and six protons is exactly 12.00000u.)

Energy is expressed in ergs (g–cm²/sec²), therefore, Δm must be converted to grams before solving for ΔE.

That is,

$$\Delta m = (-.0045 \text{ amu}) \frac{(1.673 \times 10^{-24} \text{g})}{(1.0073 \text{ amu})} = -\ 0.00747394 \times 10^{-24} \text{g}$$

$$= -7.47 \times 10^{-27} \text{g.}$$

Solving for ΔE :

$$\Delta E = (-7.47 \times 10^{-27} \text{g})\ (3.0 \times 10^{10} \text{ cm/sec})^2$$
$$= -6.72 \times 10^{-6} \text{ g cm}^2/\text{sec}^2 = -6.72 \times 10^{-6} \text{ erg}$$

The negative sign indicates that 6.72×10^{-6} ergs are released in the decay process.

Q. 14 : You add radioactive material whose activity per unit mass was 45.5 $s^{-1}g^{-1}$ to a liquid and then dilute it so the activity per unit mass was 6.5 $s^{-1}g^{-1}$. What is the dilution factor and the original amount of radioactive material? Assume 100 mg of radioactive material was added.

Solution: The dilution factor is simply the ratio of the activities since the activity is proportional to the concentration. Hence,

$$\text{dilution factor} = \frac{45.5 \text{ s}^{-1}\text{g}^{-1}}{6.5 \text{ s}^{-1}\text{g}^{-1}} = 7.0$$

Let X be the original amount of material

X* be the amount of radioactive material added

Since the activity per unit mass decreased seven times (dilution factor = 7), the mass must have increased seven times. Therefore,

$$X + X^* = 7.0X^*$$

$$X = 7.0X^* - X^* = 6.0X^* = 6.0\ (100 \text{ mg}) = 600 \text{ mg.}$$

Q. 15 : You are a chemist employed by a geological company to date the age of rocks by radioactive testing. You are told by an assistant that the sample he just handed to you contains a substantial quantity of ^{40}Ar. Through analysis you discover the sample contains 3 percent K (by weigth), that there is $\frac{95.0 \times 10^{-7} m^3\ ^{40}Ar}{kg\ sample}$ at STP, and the $^{40}Ar/^{36}Ar$ ratio is an incredible 75,200. Determine the age of the sample. How do you do it? The percent abundace of ^{40}K is 0.0118.

Solution: This problem is not a simple decay process where an element decays by one type of decay. It is actually a complex process which uses to different modes of dacy, electron capture and beta decay. Therefore, the equation used to find the time of decay is not the simple decay equation but rather a complex one of the following form.

$$t = \frac{1}{\lambda_{EC} + \lambda_{\beta-}} \ln \left[1 + \frac{(\lambda_{EC} + \lambda_{\beta-}) N_{40,Ar}}{(\lambda_{EC}) N_{40,K}} \right] \quad (1)$$

where

t = time of decay

λ_{EC} = decay constant for electron capture

$\lambda_{\beta-}$ = decay constant for beta decay

$N_{40,Ar}$ = number of atoms of ^{40}Ar

$N_{40,K}$ = number of atoms of ^{40}K

For $^{40}_{19}K$, $\lambda_{EC} = 5.85 \times 10^{-11} y^{-1}$ and $\lambda_{\beta-} = 4.72 \times 10^{-10} y^{-1}$

It is obvious from equation (1) that it is necessary to find $N_{40,Ar}$ and $N_{40,K}$. Remember, in solving this problem there will be ^{40}Ar which was formed by decay of ^{40}K, but there will also be ^{40}Ar which was not formed from decay of ^{40}K.

To calculate the number of moles of ^{40}Ar/kg sample, use the ideal gas law. Hence,

$$n = \frac{PV}{RT} = \frac{(1\ atm)(95.0 \times 10^{-7} m^3 / kg\ sample)}{\left(82.1 \frac{cm^3\ atm}{°k\ mol}\right)\left(\frac{1n}{100\ cm}\right)^3 (273° K)}$$

$$= 4.24 \times 10^{-4} mol / kg\ sample$$

The number of atoms of ^{40}Ar/kg sample can now be found by multiplying by Avogadro's number. Hence,

$N_{40,Ar}$ = 4.24×10^{-4} mol/kg sample $\times$ 6.02252×10^{23} atoms/mol

$= 2.5535 \times 10^{20}$ atoms/kg sample

As stated earlier, not all of the ^{40}Ar was derived from the ^{40}K. This must now be corrected for. In a normal sample of Ar, the ratio of $^{40}Ar/^{36}Ar$ is 296. For the sample this ratio was 75,200. This ratio for the sample represents the total amount, that is, the normal ratio plus the ratio from the decay of ^{40}K. Therefore, by subtracting the normal amount from the total, and then dividing this difference by the total, the fraction of ^{40}Ar derived from ^{40}K can be found. Hence,

$$\frac{(75,200 - 296)}{75,200} = .996$$

This is now the correction factor to be multiplied in front of $N_{40,\,Ar}$ to find $N_{40,\,Ar}$ from the decay of ^{40}K. Hence,

$$N_{40,\,Ar} = 2.5535 \times 10^{20} \text{ atoms/kg sample} \times .996$$
$$= 2.5433 \times 10^{20} \text{ atoms/kg sample}$$

Next, the number of atoms/kg sample must be found. It is known that the percent of potassium (by weigth) in 1 kg sample is 3%. Therefore, the fraction of potassium in 1000 g is $\frac{(3 \times 10^{-2})}{\text{kg sample}}(1000 \text{ g})$. Hence, the number of moles/kg sample of potassium is

$$\frac{\dfrac{(3 \times 10^{-2})\,(1000\text{g})}{\text{kg sample}}}{39.1\dfrac{\text{g}}{\text{mol}}} = .767 \; \frac{\text{mol K}}{\text{kg sample}}$$

However, not all of the potassium in the sample is ^{40}K. The percent abundance of ^{40}K is 1.18×10^{-2}. Hence, the fraction of ^{40}K in the sample is 1.18×10^{-4}. Therefore, the number of $\frac{\text{moles of}}{\text{kg sample}}$ ^{40}K is given by,

$$.767 \text{ mol/kg sample} \times 1.18 \times 10^{-4} = 9.05 \times 10^{-5} \text{ mol/kg sample}$$

The number of atoms of ^{40}K/kg sample can now be found by multiplying by Avogadro's number. Hence,

$$N_{40,\,K} = 9.05 \times 10^{-5} \text{ mol/kg} \times 6.02252 \times 10^{23} \text{ atoms/mol}$$
$$= 5.4504 \times 10^{19} \text{ atoms/kg sample}$$

All the information required in equation (1) has been found.

Therefore,

$$t = \frac{1}{\lambda_{EC} + \lambda_{\beta-}} \ln\left[1 + \frac{(\lambda_{EC} + \lambda_{\beta-}) N_{40, Ar}}{(\lambda_{EC}) N_{40, K}}\right]$$

$$= \frac{1}{5.85 \times 10^{-11} y^{-1} + 4.72 \times 10^{-10} y^{-1}}$$

ln

$$\left[1 + \frac{(5.85\times10^{-11} y^{-1} + 4.72\times10^{-10} y^{-1})\,(2.5433 \times 10^{20}\,\text{atoms/ kg sample})}{(5.85 \times 10^{-11} y^{-1})\,(5.4504 \times 10^{19}\,\text{atoms/ kg sample})}\right]$$

$$= \frac{1}{5.305 \times 10^{-10} y^{-1}}$$

$$\ln\left[1 + \frac{(5.85\times10^{-10} y^{-1})\,(2.5433 \times 10^{20}\,\text{atoms / kg sample})}{(5.85 \times 10^{-11} y^{-1})\,(5.4504 \times 10^{19}\,\text{atoms / kg sample})}\right]$$

$$= \frac{1}{5.305 \times 10^{-10} y^{-1}} \ln\left[1 + \frac{1.3492 \times 10^{11}}{3.1885 \times 10^{9}}\right]$$

$$= \frac{1}{5.305 \times 10^{-10} y^{-1}} \ln[43.32]$$

$$= 7.10 \times 10^{9} y$$

Q. 16 : ^{146}Ce decays by beta emission to ^{146}Pr, with $t_{1/2} = 14$ min; the ^{146}Pr decays in turn to stable ^{146}Nd with $t_{1/2} = 25$ min. A sample consists initially of 0.5mCi of pure ^{146}Ce. Calculate the activity of ^{146}Ce and of ^{146}Pr present 30 minutes later. Is this a case of secular equilibrium ?

Solution : The problem illustrates consecutive radioactive decay. An initial species decays to a second species which undergoes radioactive decay. The sequence of decays for this problem can be represented as

$$146_{Ce} \xrightarrow{\lambda_1} 146_{Pr} \xrightarrow{\lambda_2} 146_{Nd}$$

where λ_1, λ_2 represent the first and second decay constants respectively.

The equation dealing with such sequential decay is given by

$$D_2 = \frac{\lambda_1 D_1^0}{\lambda_2 - \lambda_1}^{\left(e^{-\lambda_1 t} - e^{-\lambda_2 t}\right)} \quad (1)$$

where

λ_1 = first decay constant

λ_2 = second decay constant

D_1^0=initial amount of the first species

D_2 = amount of the second species at some time.

t = any arbitrary time.

It is obvious that in order to calculate the activity of ^{146}Pr (the second species), it is necessary to calculate both decay constants, λ_1,λ_2. This can be done because the half-life of each decay is given. For the first decay the following equation can be written,

$$D_1 = D_1^0 e^{-\lambda_1 t} \quad (2)$$

Now the half-life is defined as the time necessary to reduce the initial concentration in half. That is, when $t = t_{1/2}$ $D_1 = \frac{1}{2} D_1^0$. Then equation (2) becomes

$$\frac{1}{2} D_1^0 = D_1^0 e^{-\lambda_1 t_{1/2}}$$

which rearranges to

$$\frac{\frac{1}{2} D_1^0}{D_1^0} = e^{-\lambda_1 t_{1/2}}$$

Taking natural logarithms of both sides and solving for λ_1, equation (3) becomes

$$\lambda_1 = \frac{.693}{t_{½}}.$$

Since $t_{½}$ = 14 min; λ_2 = 0.0495 min^{-1}.

The second decay constant, λ_2, must now be found. It does not matter that the initial concentration of the second species is unknown. At the half-life, $t_{½}$, only half the concentration remains. Hence, as was done with the first species,

$$\lambda_2 = \frac{.693}{t_{1/2}}$$

Since $t_{1/2} = 25$ min, $\lambda_2 = 0.027\ \text{min}^{-1}$.

The activity of ^{146}Ce is to be found after 30 minutes. Since the initial concentration is given, equation (2) can be used. Therefore,

$$D_1 = D_1^0 e^{-\lambda_1 t} = 0.5\ \text{mCi}\ e^{-(0.0495\ \text{min}^{-1})(30\ \text{min})}$$

$$= 0.113\ \text{mCi}$$

The activity of ^{146}Pr is to be found after 30 minutes. In this case it is necessary to use equation (1) which relates the activity of the second species to the initial activity of the first species. Therefore

$$D_2 = \frac{\lambda_1 D_1^0}{\lambda_2 - \lambda_1}\left(e^{-\lambda_1 t} - e^{-\lambda_2 t}\right)$$

$$= \frac{(0.0495\ \text{min}^{-1})\ (0.5\ \text{mCi})}{(0.027\ \text{min}^{-1} - .0495\ \text{min}^{-1})}\left(e^{-(0.0495\ \text{min}^{-1})\ (30\ \text{min})} - e^{-(.027\text{min}^{-1})\ (30\ \text{min})}\right)$$

$$= \frac{0.02475\ \text{min}^{-1}\ \text{mCi}}{-0.0225\ \text{min}^{-1}}\ (.227\ -\ .445)$$

$$= .240\ \text{mCi}$$

Secular equilibrium is when the activity of the first species does not change over several half-lives of the second species. That is, $D_2 = D_1^0$. This occurs when $\lambda_2 >> \lambda_1$.

Considering the process

$$^{146}\text{Ce} \xrightarrow{\lambda_1} {}^{146}\text{Pr} \xrightarrow{\lambda_1} {}^{146}\text{Nd}$$

if $\lambda_2 >> \lambda_1$ then ^{146}Ce has not appreciably decayed. Therefore, the activity of the first species is regarded as being constant. In this problem $\lambda_1 = 0.0495\ \text{min}^{-1}$ and $\lambda_2 = .027\ \text{min}^{-1}$. Therefore, $\lambda_1 > \lambda_2$, and hence, there is no secular equilibrium.

Q. 17 : Calculate the Coulombic barrier in the bombardment of ^{209}Bi with alpha particles.

Solution: In bombarding a nucleus with a charged particle an energy barrier is encountered. This is due to the Coulombic repulsion which develops between the charged particle and the nucleus. An empirical formula to calculate this Coulombic barrier is given by

$$E_{barrier} = \frac{0.96\ zZ}{A^{1/3}_{particle} + A^{1/3}_{target}}$$

where

z = charge number of the particle

Z = atomic number of the target

$A_{particle}$ = mass of the particle

A_{target} = mass of the target

$E_{barrier}$ = Coulombine barrier measured in MeV

For the bombardment of ^{209}Bi with alpha particles, $z = 2$, $Z = 83$, $A_{particle} = 4$ and $A_{target} = 209$. Therefore,

$$E_{barrier} = \frac{(0.96)(2)\ (83)}{(4)^{1/3} + (209)^{1/3}} = 21.2 \text{ MeV}$$

Q. 18 : Calculate the barrier potential and threshold energy for an alpha particle colliding with a ^{14}N nucleus.

Solution: In a collision of a particle with a nucleus a certain minimum energy must be imparted from the kinetic energy of the particle in order for a reaction to take place. This energy is called the threshold energy.

The following reaction is relevant to this problem.

$$^{14}_{7}N + ^{4}_{2}He \longrightarrow ^{1}_{1}H + ^{17}_{8}O$$

In order to calculate the threshold energy it is necessary to calculate the change in mass due to the collision. This is done by summing the mass of the products and summing the mass of the reactants, and then subtracting the mass of the reactants from the mass of the products. The mass of hydrogen is 1.0078252 amu and the mass of oxygen is 16.999133 amu. (Note these are the atomic masses of the specific isotopes and not the common atomic masses which are averages of all the isotopes.) therefore, the sum of the products is 18.006958 amu.

The mass of nitrogen is 14.0030744 amu and the mass of helium is 4.0026036 amu. (Note, again these are the atomic masses of the specific isotopes.) Therefore, the sum of the reactants is 18.005678 amu.

The change in the mass as a result of the collision in 18.006958 amu −18.005678 amu = 0.00128 amu. To calculate the energy equivalent of this mass change, Einstein's equation is used. It states

$$\Delta E = (\Delta m)C^2$$

where

$$\Delta E = \text{energy change}$$
$$\Delta m = \text{change in the mass}$$
$$c = \text{speed of light}$$

Note the mass in the above equation is not in dimensions of amu but rather kg (g in the cgs system). Hence,

$$\Delta E = 0.00128 \text{ amu} \times 1.66057 \times 10^{-27} \frac{\text{kg}}{\text{amu}} \times \left(2.997925 \times 10^8 \frac{\text{m}}{\text{s}}\right)^2$$

$$\times \frac{1\text{ev}}{1.602189 \times 10^{-19}\text{J}} \times \frac{1\text{MeV}}{10^6\text{eV}}$$

$$= 1.19\text{MeV} \left(\text{Remember } 1\text{J} = 1 \text{ kg m}^2/\text{s}^2\right)$$

To calculate the barrier potential an empirical formula is used. It is given by

$$E_{\text{barrier}} = \frac{0.96 \, zZ}{A_{\text{particle}}^{1/3} + A_{\text{target}}^{1/3}}$$

where

E_{barrier} = barrier potential (in MeV)

z = charge number of the particle

Z = atomic number of the target

A_{particle} = mass of the particle

A_{target} = mass of the target

For the bombardment of ^{14}N with alpha particles, $z = 2$, $Z = 7$, $A_{\text{particle}} = 4$, and $A_{\text{target}} = 14$, Therefore,

$$E_{\text{barrier}} = \frac{(0.96)(2)(7)}{(4)^{1/3} + (14)^{1/3}}$$

$$= 3.36 \text{ MeV}$$

Q. 19 : The maximum weekly dose for the hands and forearms, assuming a mass of 4kg and a surface area of 800 cm², is 1.5 rem. How long would it take for a person to receive this dose from a 10 millicurie source emitting a 2.5 MeV gamma ray? The distance between the source and the receiving surface is 10 cm. Assume no radiation is absorbed by the air.

Solution: The curie is equal to 3.700×10^{10} disintegration/sec. Therefore, the number of gamma rays emitted is equal to

$$10 \text{ millicurie} \times 10^{-3} \frac{\text{curie}}{\text{millicurie}} \times 3.7 \times 10^{10} \text{s}^{-1}\text{curie}^{-1}$$

$$= 3.7 \times 10^{8} \text{s}^{-1}$$

The next step is to calculate the flux which reaches the receiving surface. If the source is considered to be a point, then the rays move out in all directions forming a sphere. The total area covered by these rays is given by the equation for the surface area of a sphere. This is $4\pi r^2$ where r is the distance between the source and the receiving surface. Hence, the flux is

$$\frac{3.7 \times 10^{8} \text{ s}^{-1}}{4\pi \left(10\text{cm} \times 1\text{m} / 100 \text{ cm}\right)^2} = \frac{3.7 \times 10^{8} \text{ s}^{-1}}{.1257\text{m}^2}$$

$$= 2.94 \times 109 \text{ s}^{-1} \text{ m}^{-2}$$

This is the radiation flux. What is wanted is the energy flux. Therefore, this flux is multiplied by the energy of the source. Hence,

$$2.5 \text{ MeV} \times \frac{10^6 \text{eV}}{\text{MeV}} \times 1.602189 \times 10^{-19} \frac{\text{J}}{\text{eV}} \times 2.94 \times 10^9 \text{ s}^{-1}\text{m}^{-2}$$

$$= 1.18 \times 10^{-3} \text{ Js}^{-1} \text{ m}^{-2}$$

The dosage rate is found by multiplying the energy flux by the surface area of the receiving surface. Therefore,

dosage rate $= 1.18 \times 10^{-3} \text{ J s}^{-1} \text{ m}^{-2} \times 800 \text{ cm}^2 \times (1\text{m}/100 \text{ cm})^2$

$= 9.44 \times 10^{-5} \text{ J s}^{-1}$

The next step often confuses the student unfamiliar with the units of radiation. The following discussion will help. The unit called the roentgen is defined for ionizing radiation which is incident in air. It is not based on the actual amount of energy absorbed by the receiving source. It is therefore defined as the amount of ion pairs formed in air. Hence,

$1\text{r} = 16.1 \times 10^{15}$ ion pairs/kg of air,

or upon conversion,

$1\text{r} = 2.57976 \times 10^{-4}$ C/kg.

The unit of radiation which is based upon the actual energy absorbed is the rad. 1 rad $= 1.00 \times 10^{-2}$ J absorbed/kg of receiving source. The rad unit is used for all types of radiation (not just ionizing radiation). In considering damage done to living tissue, other factors besides energy release, are important. Hence each type of radiation has its relative biological effectiveness (rbe). For α particles rbe = 20, and for γ radiation, rbe = 1. The product of the dose in rads and the rbe factor is equal to the roentgen– equivalent in man, rem.

Hence, the 1.5 rem given in the problem can now be converted into energy terms.

$$\frac{1.5 \text{ rem}}{1 \text{ rbe}} = 1.5 \text{ rad and } 1.5 \text{ rad} \times 1.00 \times 10^{-2} \frac{\text{J}}{\text{kg rad}}$$

$$= 1.5 \times 10\text{–}2 \text{ J/kg}$$

Now the receiving source has a mass of 4 Kg. Therefore, the permitted dose is

$$1.5 \times 10^{-2} \text{ J/kg} \times 4 \text{ kg} = 6 \times 10^{-2} \text{ J}.$$

The time needed to achieve this dose is found by dividing the dose by the dosage rate,

$$\frac{6 \times 10^{-2} \text{J}}{9.44 \times 10^{-5} \text{J s}^{-1}} = 6.36 \times 10^{2} \text{ sec} \times \frac{1 \text{ min}}{60 \text{ sec}} = 10.6 \text{ min}.$$

Q. 20 : A chemist is given an unknown element X. He finds that the element X has an atomic weight of 210.197 amu and consists of only two isotopes, ^{210}X and ^{212}X. If the masses of these isotopes are, respectively, 209.64 and 211.66 amu, what is the relative abundance of the two isotopes ?

Solution: The relative abundance of the two isotopes is equal to their fraction in the element. The sum of the fractions of the isotopes times their respective masses is equal to the total atomic weight of X. The element X is composed of only two isotopes. The sum of their fracions must be equal to one.

Solving: Let y = the fraction of the 210 isotope. Since the sum of the fractions is one, 1–y=the fraction of the 212 isotope. The sum of the fractions times their masses, equals the atomic weight of X.

$$209.64\,y + 211.66\,(1 - y) = 210.197;$$

$$209.64\,y + 211.66 - 211.66\,y = 210.197$$

$$-2.02\,y = -1.463$$

$$y = 1.463/2.02$$

$$y = 0.7242 = \text{fraction of } ^{210}X.$$

$$\therefore 1 - y = 0.2758 = \text{fraction of } ^{212}X.$$

By percentage the relative abundance of ^{210}X is 72.42% and ^{212}X is 27.58%.

Q. 21 : Chromium exists in four isotopic forms. The atomic masses and percent occurrences of these isotopes are listed in the following table :

Isotopic mass (amu)	Precent occurrence
50	4.31%
52	83.76%
53	9.55%
54	2.38%

Calculate the average atomic mass of chromium.

Solution: By definition,

$$A = \sum_{i=1}^{N} P_i M_i, \qquad (1)$$

where A = average value, P_i = probability of occurrence of isotope i M_i = atomic mass of isotope i. Making use of the above definition and the given data, the average atomic mass of chromium can be computed.

For the four isotopes of chromium,

$M_1 = 50$ amu $P_1 = 4.31\% = 0.0431$

$M_2 = 52$ amu $P_2 = 83.76\% = 0.8376$

$M_3 = 53$ amu $P_3 = 9.55\% = 0.0955$

$M_4 = 54$ amu $P_4 = 2.38\% = 0.0238$

Hence, the average atomic mass of chromium is $A = P_1M_1 + P_2M_2 + P_2M_3 + P_4P_4 = 0.0431 \times 50$ amu $+ 0.8376 \times 52$ amu $+ 0.0955 \times 53$ amu $+ 0.0238 \times 54$ amu $= 2.155$ amu $+ 43.555$ amu $+ 5.062$ amu $+ 1.285$ amu $= 52.057$ amu.

Q. 22 : Natural chlorine is a mixture of isotopes. Determine its atomic weight if 75.53% of the naturally occurring element is chlorine 35, which has a mass of 34.968, and 24.47% is chlorine 37, which has a mass of 36.956.

Solution: If naturally occurring chlorine containe 75.53% chlorine 35 and 24.47% chlorine 37, one mole of chlorine contains .7553 mole chlorine 35 and .2447 mole chlorine 37. The weight of one mole of chlorine is, then, equal to the sum of the weight of.

.7553 mole ^{35}Cl and .2447 mole ^{37}Cl.

Making use of the definition of average value gives

$$A = P_1M_1 + P_2M_2 \quad (1)$$

where A = atomic weight

P_i = probability of occurrence of isotope i

M_i = mass of isotope i.

In this problem, $P_1 = 0.7553$, $M_1 = 34.968$

$P_2 = 0.2447$ and $M_2 = 36.956$.

Inserting these values into equation (1) yields :

$$\text{atomic weight of Cl} = (.7553)\ (34.968) + (.2447)\ (36.956)$$
$$= 26.41 + 9.04$$
$$= 35.45.$$

Q. 23 : Chlorine is found in nature in two isotopic forms, one of atomie mass 35 amu and one of atomic mass 37 amu. The average atomic mass of chlorine is 35.453 amu. What is the percent with which each of these isotopes occurs in nature ?

Solution: This is a mathematical problem, the solution to which centers on defining the average value properly. Consider a set of N observations or measurements, M_1, M_2,...,M_N. Let the probability that the observation M_1 is made be P_1, the probability that the observation M_2 is made be P_2, and so on. Then, the average, A is defined as :

$$A = \sum_{i=1}^{N} P_i M_i.$$

Since, by definition, the sum of the probabilities must be one, then :

$$1 = \sum_{i=1}^{N} P_i.$$

For this problem there are two observation, hence N = 2.

Let the first observation be M_1 = 35 amu and the second observation be M_2 = 37 amu. The average A = 35.453 amu. Hence,

$$A = P_1M_1 + P_2M_2 \quad (1)$$

$$35.453 \text{ amu} = P_1 \times 35 \text{ amu} + P_2 \times 37 \text{ amu and}$$

$$1 = P_1 + P_2. \quad (2)$$

In order to find the percent occurrences of these isotopes, the above equations are solved simultaneously to obtain the values of P_1 and P_2. Rearranging equation (2) gives :

$$P_2 = 1 - P_1.$$

Substituting this into equation (1) yields:

$$\begin{aligned} A &= P_1M_1 + P_2M_2 = P_1M_1 + (1 - P_1)\,M_2 \\ &= P_1M_1 + M_2 - P_1M_2 \\ &= M_2 + (M_1 - M_2)\,P_1 \end{aligned}$$

or

$$P_1 = \frac{A - M_2}{M_1 - M_2} = \frac{35.453\text{ amu} - 37\text{amu}}{35\text{amu} - 37} = 0.7735.$$

Then,

$$P_2 = 1 - P_1 = 1 - 0.7735 = 0.2265.$$

Thus, the isotope of mass 35 amu occurs with probability of 0.7735 or 0.7735 × 100% = 77.35% and the isotope of mass 37 amu occurs with a probability of 0.2265 or 0.2265 × 100% 22.65%.

In reality, the two isotopes do not have integral atomic masses, and the percent occurrences calculated above are not exactly correct.

Q. 24 : Compute the missing mass of $^{19}_{9}F$ (atomic weight = 18.9984) when the masses of proton, neutron and electron are 1.00728, 1.00867 and 0.000549 respectively.

Solution: The total number of particles in $^{19}_{9}F$ and their total weight can be calculated. The amount of missing mass in $^{19}_{9}F$ will be the difference of this calculated weight and the given atomic weight of $^{19}_{9}F$ The subscript number 9, in $^{19}_{9}F$, indicates the atomic number of fluorine (F). Because the atomic number equals the number of protons, there are 9 protons in F. The superscript, 19, indicates the total number of particles in the nucleus since the nucleus is composed of protons and neutrons, and there are 9 protons, there are 10 neutrons present. In a neutral atom, the number of electrons equals the number of protons. Thus, there are 9 electrons. The total number of particles is, thus, 28. The mass and quantity of each particle is now known. Calculating the total weight contribution of each type of particle :

$$\begin{aligned} \text{Protons : } 9 \times 1.00728 &= 9.06552 \\ \text{Neutrons : } 10 \times 1.00867 &= 10.0867 \\ \text{Electrons: } 9 \times 0.000549 &= \underline{0.004941} \end{aligned}$$

Total mass = 19.1572.

It is given that the mass of the fluorine atom is 18.9984. Therefore, the missing mass is 19.1572 – 18.9984 = 0.1588 amu.

Q. 25 : The natural abundance of neon is 90.92% $^{20}_{10}Ne$ (19.99244 u), 0.257% $^{21}_{10}Ne$ (20.99395 u) and 8.82% $^{22}_{10}Ne$ (21.99138 u) Calculate the average atomic weight for natural neon gas.

Solution: The average atomic weight for natural neon gas can be calculated by dividing the sum of products of the percentages times the atomic weights by the sum of the percentages.

The mathematical representation of this is

$$\frac{\sum P_i a_i}{\sum P_i} \qquad (1)$$

where P_i = percentage of the ith component and

a_i = atomic weight of the ith component.

Inserting the numerical data into equation (1) gives :

$$\frac{(0.9092)\,(19.99244) + (0.00257)\,(20.99395) + (0.0882)\,(21.99138)}{0.9092 + 0.00257 + 0.0882}$$

$$= \frac{18.177126 + 0.0539544 + 1.9396397}{0.9092 + 0.00257 + 0.0882}$$

$$= \frac{20.17072}{0.99997}$$

$$= 20.171325 \text{ u.}$$

Q. 26 : The mass of an atom of hydrogen having one proton and one electron is 1.0072766 + 0.0005486 = 1.0078252 u. What fraction of the total mass of a hydrogen atom is contained within the nucleus ? Contrast this answer to that for the fraction of the atomic volume (r = 0.529 Å) occupied by the nucleus if

$$r = \left(1.5 \times 10^{-15}\,m\right)A^{\frac{1}{3}}$$

is valid, where r = radius of a nucleus

A = mass number.

Solution: The total mass of the atom is 1.0072766 +

0.0005486 = 1.0078252 u. Observe that 1.0072766 is the mass of proton and 0.0005486 is the mass of electron.

As a result, the fraction of atomic mass is given by

$$\frac{1.0072766}{1.0078252} = 0.9994557.$$

The fraction of volume is defined as the ratio of the volume of nucleus to that of the atom. That is,

$$\frac{\left[\left(1.5 \times 10^{-5}\right)1^{\frac{1}{3}}\right]^3}{\left(0.529 \times 10^{-10}\right)^3}$$

$$= \frac{22.7985 \times 10^{-45}}{1 \times 10^{-30}}$$

$$= 22.7985 \times 10^{-15}$$

$$= 2.3 \times 10^{-14}.$$

Q. 27 : List (a) isotopes, (b) isobars and (c) isotones among the folllowing nuclides:

$${}^{15}_{8}O,\ {}^{14}_{7}N,\ {}^{13}_{6}C,\ {}^{12}_{5}B,\ {}^{14}_{8}O,\ {}^{13}_{7}N,\ {}^{15}_{7}N,\ {}^{16}_{8}O,\ {}^{18}_{9}F,\ {}^{17}_{9}F$$

Solution: Before these nuclides can be classified as either isotopes, isobars or isotones, it is important to state clearly what they mean.

Isotopes are nuclides having the same number of protons in the nucleus. That is, the atomic numbers are the same.

Isobars are nuclides having the same number of neutrons and protons in the nucleus. In other words, they have equal mass numbers and these represent the number of nucleons in the nucleus.

Isotones are nuclides having the same neutron number in the nucleus. That is, they have equal amount of neutrons in the nucleus.

(Note that a typical nuclide is represented by the symbol ${}^{m}_{p}X$ or ${}_{p}X^{m}$. m is the mass number, p is the atomic number and X is the elemental chemical symbol corresponding to the value of p.)

Now that these terminologies have been well defined, the nuclides can be categorized very easily.

(a) The isotopes are ${}^{13}_{7}N$, ${}^{14}_{7}N$ and ${}^{15}_{7}N$;

$^{14}_{8}O$, $^{15}_{8}O$, and $^{16}_{8}O$;

$^{17}_{9}F$ and $^{18}_{9}F$.

(b) The isobars are $^{13}_{6}C$ and $^{13}_{7}N$;

$^{14}_{7}N$ and $^{14}_{8}O$;

$^{15}_{7}N$ and $^{15}_{8}O$.

(c) The isotones are $^{13}_{7}N$ and $^{14}_{8}O$;

$^{12}_{5}B$, $^{13}_{6}C$, $^{14}_{7}N$ and $^{15}_{8}O$;

$^{15}_{7}N$, $^{16}_{8}O$ and $^{17}_{9}F$.

Interaction of Radiation with Matter

Q. 28 : Given that lead has a density of 11.3×10^3 kg m^{-3} and a mass absorption coefficient for 3.0 MeV, γ's corresponding to a half thickness of 16 g cm^{-2}, calculate the percent decrease in activity at a depth of 1 cm.

Solution: All the processes by which γ-rays interact with matter lead to exponential attenuation, which is characterized by a half thickness and an absorption coefficient μ.

If I_0 is the incident intensity of a beam of γ-rays, then the intensity I transmitted through a thickness x cm of absorber is given by

$$I = I_0 e^{-\mu x} \tag{1}$$

The linear absorption coefficient μ is a function of the energy of the γ-rays and of the absorbing medium. The half thickness $x_{½}$ is defined as the thickness of absorber required to absorb one half of the incident photons, that is, the thickness required to make $I = ½ I_0$. It is related to μ by the equation

$$x_{½} = \frac{0.693}{\mu}$$

The absorber thickness are freqently given in terms of surface density ρx, expresed in grams per square centimeter, where ρ = the density. Thus equation (1) can be written as

$$I = I_0 e^{-(\mu/\rho)\rho x}.$$

The ratio $\frac{\mu}{\rho}$ is called the mass absorption coefficient.

The mass absorption coefficient for this problem has a value given by

$$\frac{0.693}{16\text{g cm}^{-2}}$$

$$= \left(4.33 \times 10^{-2}\,\text{cm}^2\text{g}^{-1}\right)\left(\frac{1\text{m}}{100\text{ cm}}\right)^2\left(\frac{1000\text{g}}{1\text{kg}}\right)$$

$$= 4.33 \times 10^{-3}\,\text{m}^2\text{kg}^{-1}$$

$\therefore$ μ (linear) = μ (mass absorption) ρ

$$= \left(4.33 \times 10^{-3}\,\text{m}^2\text{kg}^{-1}\right)\left(11.3 \times 10^3\,\text{kg m}^{-3}\right)$$

$$= 48.9\text{m}^{-1}.$$

Using equation (1) yields

$$I = I_0 \exp\left(-48.9\text{m}^{-1}\right)\left(1 \times 10^{-2}\,\text{m}\right)$$

$$= 0.613\ I_0.$$

The decrease in activity at a depth of 1×10^{-2}m is therefore

$$1 - 0.613 = 0.87$$

or

$$= 38.7\%$$

Q. 29 : Determine the approximate energy-loss rate of 3 Mev alpha particles in air, given that the mean energy–loss rate of 3 Mev protons in air is 12.5×10^{-2} Mev/cm.

Solution: Particles such as protons or alpha particles, interact by virtue of their electric charge with the atomic electrons of the medium through which they are passing. As a result, part of the kinetic energy of the moving particle is transferred to the electron, thus slightly decreasing the velocity of the moving particle. The rate at which a particle loses energy along its path, –de/dx, is given by

$$-\frac{dE}{dx} = \frac{4\pi e^4 z^2}{m_0 \overline{V}^2} NZ\left[\ln\frac{2m_0\overline{V}^2}{I\left(1-\beta^2\right)} - \beta^2\right] \qquad (1)$$

where z = the charge of the moving particle whose average velocity is $\overline{V}$, m_0 and e are the rest mass and charge of an electron respectively. N is the number of stopping atoms per cm^3, and Z is their atomic number. β is the ratio $\overline{V}/c$, where c = the velocity of light. I is the average energy required to excite of ionize the atoms of the stopping material.

For particles whose velocities are much smaller than c, the terms containing β may be ignored. Also the logarithmic term will vary rather slowly with $\overline{V}$. Thus equation (1) may be written in an approximate form as

$$-\frac{dE}{dx} \propto \frac{MZ^2}{E} \quad (2)$$

where M and E = and mass and energy of the moving particle respectively. The α is a proportionality sign. Equation (2) can also be written as

$$-\frac{dE}{dx} = \frac{4\pi Z^2 e^4 N}{m_0 \overline{V}^2}. \quad (3)$$

Since the range in a particular stopping medium is known, the range of another particle of different mass and charge can be calculated by using the subscript 1 for protons and 2 for alphas. According to equation (3),

$$\frac{-dE_1/dx_1}{dE_2/dx_2} = \frac{4\pi Z_{11}^2 e^4 N_1}{m_0 \overline{V}_1^2} \; \frac{m_0 \overline{V}_2^2}{4\pi Z_{12}^2 e^4 N_2}$$

$$= \frac{Z_{12}^2 N_1 \overline{V}_2^2}{Z_{12}^2 N_2 \overline{V}_1^2} \quad (4)$$

where $Z_{11} = 1$ and $Z_{12} = 2$. Since $E = m\overline{V}^2$, $E_1 = 3$Mev, $E_2 = 3$Mev, $m_1 = 1$ and $m_2 = 4$, $E_1 = 3 = 1\overline{V}_1^2$. Therefore, $\overline{V}_1^2 = 3$. Also $E_2 = 3 = 4\overline{V}_2^2$. Therefore $\overline{V}_2^2 = 0.75$. Using equation (4) gives-

$$\frac{-dE_2}{dx_2} = \left(12.5 \times 10^{-2}\right) \frac{(4)(3)}{(1)(0.75)}$$

$$= 2 \text{ Mev/cm}$$

Q. 30 : Calculate the ratio of energy loss when a 4 Mev alpha particle travelling through carbon collides with nuclei (scattering) to the energy loss when it collides with bound electrons.

Solution: If a particle of energy E and momentum $p = m\overline{V}$ is deflected through an angle θ, the nonrelativistic energy loss in the forward direction is equal to

$$E = \frac{m(\overline{V}\sin\theta)^2}{2} = \frac{p^2 \sin^2\theta}{2m} \tag{1}$$

The probaility $P(\theta)\,d(\theta)$ of an incident particle's being scattered through an angle between θ and $d\theta$ is given by

$$p(\theta)\,d(\theta) = \frac{\pi Z_1^2 Z_2^2 e^4 \sin\theta}{8E^2 \sin^4\theta/2} \tag{2}$$

The energy loss, $\Delta\overline{E}$, averaged over all deflections is the probability of the energy loss, Eq. (2), times the energy loss,

$$\Delta\overline{E} = \int \Delta E(\theta)\, P(\theta)\, d\theta$$

$$= \frac{p^2}{2m}\,\frac{\pi Z_1^2 Z_2^2 e^4}{8E^2} \int \frac{\sin^3\theta}{\sin^4\frac{\theta}{2}}\, d\theta \tag{3}$$

Equation (3) is integrated with the upper limit of $\theta = \pi$, which is the maximum angle when the incident particle is deflected 180°. The lower limit will be designated by θ_{min} for purposes of integration.

Integrating Eq. (3) between these limits gives

$$-\Delta E = \frac{p^2}{2m}\,\frac{\pi Z_1^2 Z_2^2 e^4}{E^2}\,2\left[\ln\left(\frac{2}{\theta_{min}}\right) + \frac{\theta_{min}^2}{4} - 1\right] \tag{4}$$

The minimum angle, θ_{min}, is related to the closest distance to which a particle of velocity $\overline{V}$ can approach a nucleus of charge Z_2. Experimental and theoretical evidence indicates that this distance is a function of the size and charge of the nucleus as well as of the velocity and charge of the incident particle. In general, the distance approximates the size of the nucleus (2×10^{-12} cm), and the angle θ_{min} for light elements is about 10^{-6} radians. The term $\theta_{min}^2/4$ and the –1 in the brackets of Eq. (4) may be neglected in comparison to the In ($2/\theta_{min}$) term.

$$-\Delta E = \frac{p^2}{2m}\,\frac{\pi Z_1^2 Z_2^2 e^4}{E^2}\,2\left[\ln\left(\frac{2}{\theta_{min}}\right)\right] \tag{5}$$

The energy loss due to scattering over an interval dx is obtained by multiplying the right-hand side of equation (5) by N_1 dx, where N_1 = the number of particles per cubic centimeter and E is substituted for $P^2/2m$ to give

$$-\frac{dE}{dx} = \frac{\pi Z_1^2 Z_2^2 e^4}{E^2} \ln(2/\theta_{min}) \qquad (6)$$

The ratio of scattering losses to collision losses, dE_s/dE_c, is

$$\frac{-dE_s/dx}{-dE_c/dx} = \frac{\frac{2\pi Z_1^2 Z_2^2 e^4 N_1}{E} \ln(2/\theta_{min})}{\frac{4\pi Z_1^2 e^4 N}{m_0 V^2} \ln\left(\frac{2m_0 \overline{V}^2}{I}\right)}$$

Substituting the following values: $\theta_{min} = 8 \times 10^{-5}$, $Z_1 = 2$, $I \cong 80$, $\overline{V} = 1.40 \times 10^9$ cm/sec, $m_0 = 9.11 \times 10^{-28}$g, $Z_2 = 6$, $N = 3.01 \times 10^{23}$ electrons/g, $N_1 = 0.5 \times 10^{-23}$ nuclei/g, $E = 4$, and $2m_0\overline{V}^2 = (2)(9.11 \times 10^{-28})(1.40 \times 10^9)^2 (6.24 \times 10^5) = 22.3 \times 10^{-4}$ Mev, and cancelling the Z_1, π, and e^4 terms, gives

$$\frac{dE_s}{dE_c} = \frac{(2)(6)^2(0.5)(10^{-23})/4 \ \ln(2/(8 \times 10^{-5}))}{(4)(3.01)(10^{-23})/(11.15 \times 10^{-4}) \ \ln(2230.80)}$$

$$= 2.5 \times 10^{-3}$$

Q. 31 : Calculate the total and average binding energies and the packing fraction for $^{12}_{6}C$.

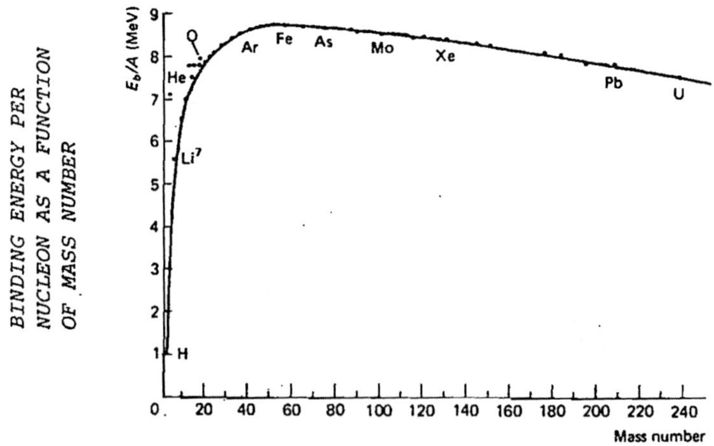

Solution: The general form of representing a particular nuclear reaction is given by

$$^{A}_{Z}X = Z\,^{1}_{1}H + (A-Z)\,^{1}_{0}n \qquad (1)$$

where $X =$ elemental chemical symbol corresponding to the value of Z.

$Z =$ atomic number

$A =$ protons + neutrons = mass number.

Consequently, for $^{12}_{6}C$, the reaction is

$$^{12}_{6}C = 6\ ^{1}_{1}H + 6\ ^{1}_{0}n. \qquad (2)$$

For a nucleus, the average binding energy is defined as the total binding energy divided by the value of A. The total binding energy is the energy for reaction (1) as calculated from the difference in the rest masses of products and reactants. Therefore, the average binding energy for reaction (2) is given by

$$\frac{\Delta(\text{mass})}{12} = \frac{6(1.007825) + 6\ (1.008665) - 12.000000}{12}$$

$$= \frac{0.098940\text{u}}{12}$$

$1\text{u} = 1$ amu, $\Delta E = (\Delta m)c^2$ where c is the speed of light. Therefore,

$$\frac{0.09894\ \text{amu} \times 1.6605(10^{-24})\text{g amu}^{-1} \times (2.997925(10^{10})\text{cm sec}^{-1})^2 \times 10^{-7}\text{J sec}^{-1}}{1.6022(10^{-9})\text{J ev}^{-1} \times 10^{6}\text{ev Mev}^{-1}}$$

$$= 92.163 \text{ Mev}$$

$$\text{Thus,}\ \frac{\Delta(\text{mass})}{12} = \frac{92.163 \text{ Mev}}{12}$$

$$= 7.680 \text{ Mev}$$

The packing fraction is given by

$$\text{pf} = \frac{(M - A)10^4}{A} \qquad (3)$$

where pf = packing fraction. M – A mass excess = mass of the nuclide in u less the mass number. In this problem, observe that M = A. As a result, equation (3) yields

$$\text{pf} = \frac{(12 - 12)10^4}{12}$$

$$= \frac{(0)10^4}{12}$$

or $\quad pf = 0$

Q. 32 : Calculate the recoil energy of an atom of mass 10, following emission of a 2 Mev gamma ray.

Solution: The energy equivalent to one mass unit is $E = mc^2 = 931 \times 10^6$ ev.

$$E_r = \frac{(2\times10^6)^2}{1862\times10^6\times10} = 214 \text{ ev},$$

where E_r = recoil energy

or, alternatively,

$$E_r = \frac{E_\gamma^2}{2\,Mc^2}$$

$$= \frac{4\times10^{12}(\text{ev})^2}{2\times10\dfrac{\text{g}}{\text{mole}}\times\dfrac{\text{mole}}{6.02\times10^{23}\text{ molec}}\times9\times10^{20}\dfrac{\text{cm}^2}{\text{sec}^2}}$$

$$= \frac{4\times10^{12}(\text{ev})^2}{\dfrac{180\times10^{20}\text{ ergs}}{6\times10^{23}\text{ molecules}}\times\dfrac{\text{ev}}{1.6\times10^{-12}\text{ ergs}}}$$

$$= \frac{4\times10^{12}\text{ ev}}{186.2\times10^8\text{ recoil}}$$

$$E_r = 214 \text{ ev / recoil}$$

Q. 33 : A sheet of gold 0.3 mm thick and 5cm^2 in area with a density of 19.3g cm^{-3} is exposed to a thermal neutron flux of 10^7 neutrons per cm^2 per second. Given that the capture cross section of ^{197}Au for thermal neutrons is 99 barns, calculate the number of radioactive ^{198}Au nuclei produced per second in the gold. Neglect any other neutron reaction with gold and take the atomic weight of gold to be 197.2.

Solution: The probability of a nuclear process is generally expressed in terms of a cross section, σ which has the dimensions of area. For a beam of particles striking a thin target, that is, a target in which the beam is slowed down only infinitesimally, the cross section for a particular process is given by

$$R_i = I\,n\,\sigma_i x \qquad (1)$$

where R_i = the number of processes of the type under consideration occurring in the target per unit time, I is the number of incident particles per unit time, n is the number of target nuclei per cubic centimeter of target, σ_i is the cross section for the processes and x is the target thickness in cm. If instead of a thin target, there is a thick target where the incident particle beam is slowed down or attenuated, the attenuation –dI in the infinitesimal thickness dx is given by

$$-dI = In\sigma_t\, dx \qquad (2)$$

Here σ_t is the total cross section. Integrating equation (2) gives

$$I = I_0 e^{-n\sigma_i x}$$

and

$$I_0 - I = I_0\left(1 - e^{-n\sigma_i x}\right) \qquad (3)$$

where I = the intensity of the beam after traversing a target thickness x, I_0 is the incident intensity, and $I_0 - I$ is the number of reactions per unit time which is what this problem requires. Since the density of gold is 19.3g cm^{-3} and its atomic weight is 197.2, therefore

$$n = \frac{19.3\text{g cm}^{-3}}{197.2\text{g mole}^{-1}} \times 6.02 \times 10^{23}\,\text{mol}^{-1}$$

$$= 5.89 \times 10^{22}\ \text{nuclei / cm}^3$$

$$x = 0.03\ \text{cm}$$

$$I_0 = \left(10^7 \frac{\text{neutrons}}{\text{cm}^2\ \text{sec}}\right)\left(5\ \text{cm}^2\right)$$

$$= 5 \times 10^7\,\text{neutrons per sec.}$$

Using these values in equation (3) yields

$$I_0 - I = 5 \times 10^7\left(1 - e^{-5.89 \times 10^{22} \times 99 \times 10^{-24} \times 0.03}\right)$$

$$= 5 \times 10^7(0.1604)$$

$$= 8.02 \times 10^6\ {}^{198}\text{Au nuclei formed per second.}$$

Q. 34 : Caculate ΔE for the proposed basis of an absolutely clean source of nuclear energy

$${}^{11}_{5}\text{B} + {}^{1}_{1}\text{H} \rightarrow {}^{12}_{6}\text{C} \rightarrow 3\ {}^{4}_{2}\text{He}$$

Atomic masses : ^{11}B = 11.00931, ^{4}He = 4.00260, ^{1}H = 1.00783.

Solution: Then energy of nuclear reactions, ΔE, is calculated from the difference between the masses of products and reactants in accordance with the Einstein Law. Einstein's Law can be stated as $\Delta E = \Delta mc^2$, where Δm = the difference in the masses of the products and the reactants, and c is the speed of light (3×10^{10} cm/sec). The total reaction here can be written as

$$^{11}_{5}B + ^{1}_{1}H \rightarrow 3\ ^{4}_{2}He$$

Δm is equal to the mass of $^{11}_{5}B$ and $^{1}_{1}H$ subtracted from the mass of 3 $^{4}_{2}He$. The mass of 3 $^{4}_{2}He$ is equal to 3 times the mass of $^{4}_{2}He$.

Thus, $$\Delta m = (3 \times m \text{ of } ^{4}_{2}He) - (m \text{ of } ^{11}_{5}B + m \text{ of } ^{1}_{1}H)$$
$$= (3 \times 4.00260) - (11.00931 + 1.00783)$$
$$= 12.0078 - 12.01714 = -9.34 \times 10^{-3}\ g/mole.$$

Now, using Δm to solve for ΔE yields

$$\Delta E = \Delta mc^2 = -9.34 \times 10^{-3}\ g/mole \times (3.0 \times 10^{10}\ cm/sec)^2$$
$$= -9.34 \times 10^{-3}\ g/mole \times 9.0 \times 10^{20}\ cm^2/sec^2$$
$$= -8.406 \times 10^{18}\ g\ cm^2/mole\ sec^2$$
$$= -8.406 \times 10^{18}\ ergs/mole.$$

There are 4.18×10^{10} ergs in 1 kcal, thus ergs can be converted to kcal by dividing the number of ergs by the conversion factor, 4.18×10^{10} ergs/kcal.

$$\text{no. of kcal} = \frac{-8.406 \times 10^{18}\ ergs/mole}{4.18 \times 10^{10}\ ergs/kcal}$$
$$= -2.01 \times 10^{8}\ kcal/mole.$$

ΔE for this reaction is -2.01×10^{8} kcal per mole.

Q. 35 : Calculate the Q value of the following reaction :

$$^{106}_{46}Pd + ^{1}_{0}n \rightarrow ^{106}_{45}Rh + ^{1}_{1}H + Q$$
$$^{106}_{45}Rh \rightarrow ^{106}_{46}Pd + \beta^{-} + \bar{\nu} + 3.54\ MeV$$

The product $^{106}_{45}Rh$ decays with a half-life of 30 seconds and

3.54 MeVβ particles are emitted to the ground state of ^{106}Pd.

Solution: Just like chemical reactions, nuclear reactions are accompanied by energy changes. Some reactions are exothermic, and some are endothermic. When energy is evolved, it appears as kinetic of the escaping particles and of the product nucleus. The change in kinetic energy in a nuclear reaction is called the Q value. Q values may be calculated from atomic masses, if they are known; but if they are not known it is often possible to calculate the Q value provided that the product nucleus is radioactive and decays back to the initial nucleus with known decay energy.

$$^{106}_{46}Pd + ^{1}_{0}n \rightarrow ^{106}_{45}Rh + ^{1}_{1}H + Q \qquad (1)$$

$$^{106}_{45}Rh \rightarrow ^{106}_{46}Pd + \beta^- + \overline{\nu} + 3.54 \text{ MeV} \qquad (2)$$

Here β^- is an electron and $\overline{\nu}$ is an antineutrino.

Adding equations (1) and (2) gives

$$^{1}_{0}n \rightarrow ^{1}_{1}H + \beta^- + \overline{\nu} + Q + 3.54 \text{ MeV.}$$

This is equivalent to the transformation of a neutron into a proton, an electron, together with an antineutrino and an energy change. Writing an energy balance on the net change gives

$$M_n \rightarrow M_{^1H} + Q + 3.54 \text{ MeV} \qquad (3)$$

where M_n = mass of neutron and $M_{^1H}$ is mass of proton

$M_n = 1.008665$ mass units

$M_{^1H} = 1.007825$ mass units

Therefore, using equation (3) yields

$$Q \text{ value} = (M_n - M_{1H}) - 3.54 \text{ MeV}$$

$$= \left[(1.008665 - 1.007825)\text{amu} \times \left(\frac{931.5 \text{ MeV}}{1 \text{ amu}}\right)\right] - 3.54 \text{ MeV}$$

$$= (0.7824 - 3.54)\text{MeV}$$

$$= -2.76 \text{ MeV.}$$

13
Stoichiometry Problems

Q. 1 : The compound calcium cyanamide, $CaCN_2$, is prepared in considerable amounts for use as a solid fertilizer. The solid mixed with the water in the soil slowly adds ammonia and $CaCO_3$ to the soil.

$$CaCN_2 + 3H_2O \rightarrow CaCO_3 + 2NH_3$$

What weight of ammonia is produced as 200 kg of $CaCN_2$ reacts ?

Solution : From the stoichiometry of this equation, 1 mole of $CaCN_2$ produces 2 moles of ammonia. The number of moles of $CaCN_2$ (MW = 80 g/mole) in 200 kg is

$$\frac{200{,}000 \text{ g } CaCN_2}{80 \text{ g / mole}} = 2{,}500 \text{ moles } CaCN_2$$

Thus, (2,500 × 2) = 5,000 moles of ammonia is produced and the weight of ammonia is its molecular weight (MW = 17 g/mole) times the number of moles produced:

(17 g/mole) (5000 moles) = 85,000 g = 85 kg of ammonia produced.

Q. 2 : A tomato weighing 60.0 g is analyzed for sulfur content by digestion in concentrated HNO_3 followed by precipitation with Ba to form $BaSO_4$. 0.466 g of $BaSO_4$ is obtained. If all of the sulfur is converted to $BaSO_4$, what is the percentage of sulfur in the tomato?

Solution : The percentage of sulfur in the tomato is the weight of sulfur in the tomato divided by the total weight of the tomato multiplied by a factor of 100. To find the percentage, the weight of sulfur must be calculated. This is done by claculating the amount of sulfur precipitated out or, in other words, the amount of sulfur present in 0.466 g of $BaSO_4$.

Using the Law of Definite Proportions, 32 g (1 mole) of sulfur are consumed to form 233 g (1 mole) of $BasO_4$. Thus, once the percentage by weight of S in $BaSO_4$ is determined, then the weight of sulfur can be calculated.

The fraction of S in $BaSO_4$ is

$$\frac{\text{MW of S}}{\text{MW of BaSO}_4} = \frac{32\ \text{g / mole}}{233\ \text{g / mole}} = 0.14$$

The weight of sulfur in 0.466 g of $BaSO_4$ is thus,

(0.14) (0.466 g) = 0.065 a.

Now, that the weight of sulfur is knwon, the percentage of sulfur in the tomato is

$$\frac{0.065\ \text{g of S}}{60.0\ \text{g of tomato}} \times 100 = 0.11\%$$

Q. 3 : Solder is a lead-tin alloy. 1.00 gram of the alloy treated with warm 30 % HNO_3 gave a lead nitrate solution and some insoluble white residue. After drying, the white residue weighed 0.600 g and was found to be SnO_2. What was the percentage tin in the sample?

$$Sn + 4HNO_3 \rightarrow SnO_2 + 2H_2O$$

Solution : From the stoichiometry of this equation, 1 mole of Sn produces **1 mole of SnO_2.** Thus, if the number of moles of SnO_2 is known, the number of moles of Sn is also known (it is equivalent to the number of moles of SnO_2). Once the number of moles of Sn is known, the weight of Sn can be calculated and thus the percentage of Sn by weight in 1.00 gram of solder alloy can be determined.

The number of moles of SnO_2 (MW = 151 g/mole) is

$$\frac{0.600\ \text{g of SnO}_2}{151\ \text{g / mole}} = 0.00397\ \text{moles.}$$

The weight of 0.00397 moles of Sn (at. wt. 119 g/mole) is (0.00397 moles) (119 g/mole) = 0.427 g. Thus, the percentage of Sn in the alloy is

$$\frac{0.472\ \text{g Sn}}{1.00\ \text{g alloy}} \times 100 = 47.2\%.$$

Q. 4 : 0.20 g of cuttings from a silver coin (containing Cu) was completely dissolved in HNO_3. Then a NaCl solution was added to precipitate AgCl, which, when collected and dried, weighted 0.2152g. What is the indicated percent of Ag in the coin ?

Solution : To solve this problem, we must (1) set up the reactions (2) determine the number of moles AgCl precipitated (3) determine the number of moles of Ag used from the stoichiometry of the equations (4) calculate the weight of Ag used (5) calculate the percent of Ag in the coin.

The equations are:

(i) $Ag + Cu + HNO_3 \rightarrow AgNO_3 + Cu(NO_3)2 \downarrow$

(ii) $AgNO_3 + NaCl \rightarrow AgCl \downarrow + NaNO_3$.

The stoichiometry of these equations says that 1 mole of AgCl is produced from 1 mole of Ag metal. From the data given, the AgCl weighed 0.2152 g. Therefore, the number of moles of AgCl precipitated can be calculated, since the molecular weight of AgCl is 143.4 g/mole. Therefore, $\frac{.2152 \text{ g}}{143.4 \text{ g / mole}} = 1.50 \times 10^{-3}$ moles of AgCl precipitated out.

Thus, 1.50×10^{-3} moles of Ag metal are consumed. The weight of Ag metal is the number of moles of Ag times its atomic weight (At. wt. = 107.868 g/mole), which is equal to (1.50×10^{-3} moles) (107.868 g/mole) = 0.162 g.

The percent of Ag metal in the coin is $\frac{0.162 \text{ g Ag}}{0.20 \text{ g}} \times 100 = 81\%$.

Q. 5 : The molecular weight of nicotine, a colorless oil, is 162.1 and it contains 74.0% carbon, 8.7% hydrogen, and 17.3 % nitrogen. Using three significant figures. for the atomic weights, calculate the molecular formula of nicotine.

Solution : A molecular formula gives the actual compoistion in number of atoms per molecule. It also summarizes the weight composition of the substance. Hence, to calculate such a formula, the molecular wieght, the weight compositon of the substance, and the atomic weights of the constitutent elements must be known.

The weight composition of the substance in nicotine, assuming there are 100 g of it, are

C = 74.0 g, H = 8.7 g, N = (110 –74.0 – 8.7) = 17.3 g.

The atomic weights of these substances are C = 12.0 g/mole, H = 1.01 g/mole, N = 14.0 g/mole.

The number of moles of each constituent is, therefore,

$$\text{moles C} = \frac{74.0 \text{ g}}{12.0 \text{ g / mole}} = 6.17 \text{ moles}$$

$$\text{moles H} = \frac{8.7 \text{ g}}{1.01 \text{ g / mole}} = 8.61 \text{ moles}$$

$$\text{moles N} = \frac{17.3 \text{ g}}{14.0 \text{ g / mole}} = 1.24 \text{ moles.}$$

Thus, the mole ratios of C : H : N are 6.17 : 8.61 : 1.224. The empirical formula is

$$C_{6.17}H_{8.61}N_{1.24} \text{ or } C_{\frac{6.17}{1.24}}H_{\frac{8.61}{1.24}}N_{\frac{1.24}{1.24}} = C_5H_7N_1.$$

To determine the molecular formula, which is related to the formula weight of the empirical formula by a whole number, divide the molecular weight (MW Nicotine = 162.1) by the formula weight and multiply this whole number times the ratio of atoms in the empirical formula. Thus,

the formula weight of $C_5H_7N_1$ = 81.05

the molecular weight of nicotine = 162.1

$$\text{the whole number} = \frac{162.1}{81.05} = 2.$$

The molecular formula is $(C_5H_7N_1)_2 = C_{10}H_{14}N_2$.

Q. 6 : The analysis of a smaple of soil weighing 0.4210 g yields a mixture of KCl and NaCl weighting 0.0699 g. From the KCl, 0.060 g AgCl is precipitated out. Calculate the percent of KCl in the soil.

Solution : To solve this problem

(1) set up the equation for the reaction of KCl to AgCl

(2) determine the number of moles of AgCl precipitated

(3) from the stoichiometry of equation determine how many moles of KCl were consumed

(4) calculate the weight of KCl consumed

(5) calculate the percentage of KCl in the soil by dividing the

weight of KCl by the total weight of soil and multiplying by 100.

The equation for the reaction is

$KCl + Ag^+ \rightarrow AgCl + K^+$

The number of moles of AgCl is

$$\frac{\text{weight of AgCl}}{\text{Mw of AgCl}} = \frac{0.60 \text{ g}}{143.5 \text{ g/mol}} = 4.18 \times 10^{-4} \text{ moles}$$

From the stoichiometry of the equation, 4.18×10^{-4} moles of KCl is consumed.

The weight of KCl consumed is the number of moles of KCl times its molecular weight. Thus,

weight of KCl = (4.18×10^{-4} moles) (74.6 g/mole) = 0.0312 g

The percentage of KCl in the soil is

$$\frac{\text{weight of KCl}}{\text{weight of soil}} \times 100 = \frac{0.0312 \text{ g KCl}}{0.4210 \text{ g Soil}} = 7.41\%.$$

Q. 7 : When ammonium nitrate (NH_4NO_3) is heated, laughing gas (N_2O) and water are produced. If the laughing gas is collected at STP (Standard Temperature and Pressure) and one starts with 10 grams of NH_4NO_3, how many liters of laughing gas could be collected ?

Solution : When any gas is collected at STP (0°C and 1 atm), one mole of it will occupy 22.4 liters. The balanced equation for this chemical reaction is

$$NH_4NO_3 \rightarrow N_2O + 2H_2O.$$

One mole of NH_4NO_3 yields one mole of N_2O. If the number of moles of NH_4NO_3 that reacted is known, the number of moles of laughing gas produced is also known.

The molecular weight of ammonium nitrate = 80 g/mole. 10 g. of ammonium nitrate (10 g/ 80 g/mole) = 0.125 moles of NH_4NO_3 produce 0.125 moles of laughing gas.

Recall that 1 mole of any gas (at STP) occupies 22.4 liters. To find out what volume 0.125 moles occupy, set up the following proportion :

$$\frac{1 \text{ mole}}{22.4 \text{ liters}} = \frac{0.125}{x}$$, where x = the volume. Solving,

x = 2.8 liters = volume of laughing gas collected.

Q. 8 : Solid ammonium nitrate, amitol, is much used as an explosive; it requires a detonator explosive to cause its decomposition to hot gases.

$$NH_4NO_3 \rightarrow N_2O + 2H_2O$$

Assume 900 g NH_4NO_3 almost instantaneously decomposed. What total volume of gas in liters forms at 1092°C ?

Solution : From the stoichiometry of the equation, 1 mole of NH_4NO_3 produces a total of 3 moles of products (1 mole of N_2O + 2 moles of H_2O).

The number of moles of NH_4NO_3 900 g (MW = 80 g/mole) is

$$\frac{900 \text{ g}}{80 \text{ g/mole}} = 11.25 \text{ moles of } NH_4NO_3.$$

There are, thus, a total of 3 × 11.25 moles = 33.75 moles of products. Using the relationship that 1 mole of any ideal gas has a volume of 22.4 ℓ at STP (Standard Temperature and Pressure), the total volume of gaseous products is

$$(22.4\ \ell/\text{mole})\ (33.75 \text{ moles}) = 756\ \ell \text{ at STP.}$$

To obtain the volume at 1092°C, Charles' Law is used

$$\frac{V_1}{V_2} = \frac{T_1}{T_2}$$

where $V_1 = 756\ \ell$, $T_1 = 273°K$ (the STP temperature), and T_2 = 1092° + 273° = 1365°K. Solving for V_2,

$$V_2 = \frac{V_1T_2}{T_1} = \frac{(756\ \ell)\ (1356°K)}{(273°K)} = 3780\ \ell.$$

Q. 9 : Lunar soil contains ilmenite, a mineral with the composition $FeO.TiO_2$. It has been suggested that moon explorers might obtain water and oxygen from lunar soil by reducing it with hydrogen (brought from earth) according to the following reactions. Beginning with soil heated to 1300°C in a solar furnace:

$$FeTiO_3 + H_2 \longrightarrow H_2O + Fe + TiO_2$$

$$2H_2O \xrightarrow{\text{electrolysis}} 2H_2 + O_2$$

How much water and oxygen could be obtained from 50 kg of soil if the soil is taken to be 5 per cent ilmenite ?

Solution : From the stoichiometry of these equations, 1 mole of $FeO.TiO_2$ is consumed to form 1 mole of H_2O and 2 moles of H_2O is consumed to form 1 mole of O_2.

If 5% (0.05) of the soil is $FeO.TiO_2$ then only

(50,000 g) (0.05) = 2,500 grams are ilmenite.

The molecular weight of ilmenite is 152 g/mole. Thus, the number of moles of ilmenite is

$$\frac{2{,}550 \text{ g ilmenite}}{152 \text{ g/ mole ilmenite}} = 16.4 \text{ moles.}$$

Thus, there must be 16.4 moles H_2O and the weight of H_2O is

(16.4 moles) (18.02 g/mole) = 296 g H_2O.

If there are 16.4 moles of H_2O, then there are 8.2 (or half 16.4 moles) moles of O_2. Thus,

$$(8.2 \text{ moles}) (32 \text{ g/mole}) = 262 \text{ g } O_2.$$

Q. 10 : The equation for cream of tartar and baking powder, in action with hot water is :

$$NaHCO_3 + KHC_4C_6O_6 \rightarrow NaKC_4H_4O_6 + H_2O + CO_2(g)$$

From the use of 4.2 g of $NaHCO_3$ (baking powder), what volume of CO_2 should be liberated through the dough at 273°C ?

Solution : From the stoichiometry of the equation, 1 mole of $NaHCO_3$ produces 1 mole of CO_2. The molecular weight of $NaHCO_3$ is 84 g/mole. Thus, 4.2g is (4.2 g/84 g/mole) 0.05 moles. 1 mole of any ideal gas at STP (Standard Temperature and Pressure) occupies 22.4 ℓ; thus the volume of CO_2 gas liberated at STP would be

$$(22.4\ \ell/\text{mole}) (0.05 \text{ mole}) = 1.12\ \ell.$$

However, this must be converted from 1.12 ℓ at STP (0°C = 273°K, 1 atm) to a temperature of 273°C = 546°K.

To this conversion Charles' Law is used

$$\frac{T_1}{T_2} = \frac{V_1}{V_2},$$

where T is temperature in °K, and V is the volume at that temperature. Thus,

$$\frac{546^\circ K}{273^\circ K} = \frac{V_1}{1.12\ \ell}$$

Volume of $CO_2 = V_1 = 2.24\ \ell$.

Q. 11 : Sea shells are mostly calcium carbonate. When heated, calcium carbonate decomposes into calcium oxide (lime) and carbon dioxide. (a) Write the balanced equation for the reaction. (b) Calculate the number of kilograms of calcium carbonate needed to produce 10 kg of lime.

Solution : Calcium carbonate decomposes as follows :

$$CaCO_3 \rightarrow CaO + CO_2$$

From the stoichiometry of this equation, 1 mole of $CaCO_3$ decomposes into 1 mole of CaO and 1 mole of CO_2. If the number of moles of lime is known, the number of moles of $CaCO_3$ can be determined. The number of moles of lime is its weight divided by its molecular weight (56 g/mole) or

$$\text{number of moles of lime} = \frac{10{,}000\ g}{56\ g/mole} = 179 \text{ moles.}$$

Since 179 moles of lime are produces, 179 moles of $CaCO_3$ are consumed. The weight of 179 moles of $CaCO_3$ is the weight of one mole of $CaCO_3$ (or the molecular weight). Thus,

the weight of $CaCO_3$ = (179 moles) (100 g/mole) = 17,900 g or 17.9 kg of $CaCO_3$ consumed.

Q. 12 : Write equations to show how $Ca(HCO_3)_2$ is involved in (a) stalagmite formation in caves, (b) formation of scale in tubes or pipes carrying hot water, (c) the reaction with soap ($C_{17}H_{35}COONa$), and (d) reaction with Na_2CO_3 in water softening.

Solution : (a) To answer this, one must know the composition of stalagmites, which are rocks composed of $CaCO_3$ and are formed in underground caverns. Thus, one knows that somehow the reaction $Ca(HCO_3)_2 \rightarrow CaCO_3$ occurs.

Since Ca^{+2}, CO_3^{-2} are not reduced or oxidized, no electrons are exhanged.

One needs to balance the reaction with respect to mass.

$$Ca(HCO_3)_2 \rightarrow CaCO_3 + H_2O + CO_2$$

This equation has both charge and balanced.

(b) The formation of scales in pipes is the result of $CaCO_3$ build up from the decomposition of bicarbonates. The reaction for this process is

$$Ca\,(HCO_3)_2 \rightarrow CaCO_3 + H_2O + CO_2$$

(c) The reaction of Ca $(HCO_3)_2$, with soap ($C_{17}H_{35}COONa$)

occurs when the Na^+ on the soap molecule is replaced by the Ca^{++}.

$$Ca(HCO_3)_2 + 2C_{17}H_{35}COONa \rightarrow (C_{17}H_{35}COO)_2Ca + H_2O + CO_2$$

(d) Permanent hardness in water results from chlorides or sulfates of metals, such as calcium or magnesium. Boiling does not remove these salts; a substance called water softener must be added to the water to form a precipitate with the salts. Water containing Ca $(HCO_3)_2$ can be softened by adding the proper amount of Na_2CO_3. The reaction is $Ca(HCO_3)_2 + Na_2CO_3 \rightarrow CaCO_3(s) + 2NaHCO_3$.

Q. 13 : A red signal flare for use on railroads is made by mixing strontium nitrate, carbon, and sulfur in the proportions as shown by the equation

$$2Sr(NO_3)_2 + 3C + 2S \rightarrow$$

Complete and balance the equation (no atmospheric oxygen is necessary).

Q. 14 : One problem concerning space flight is the removal of carbon dioxide (CO_2, molecular weight = 44.0 g/mole) emitted by the human body (about 924 g of CO_2 per day). One solution is to react CO_2 with sodium hydroxide (NaOH, molecular weight = 40.0 g/mole) to form water and sodium carbonate (Na_2CO_3, molecular weight = 94 g/mole) according to the following reaction :

$$2NaOH + CO_2 \rightarrow Na_2CO_3 + H_2O.$$

How much NaOH must be carried on board a space capsule to remove the CO_2 produced by an astronaut on a 10 day flight ?

Solution : Two moles of NaOH are required to remove one mole of CO_2, as indicated by the coefficients in the reactions. The number of moles of CO_2 produced is calculated from the mass emitted by the astronaut during the flight. From this, the number of moles of NaOH required for the removal of the CO_2 is calculated, and this number of moles is then converted to grams.

At the rate of 924 g per day, in 10 days the astronaut will produce 924 g/day × 10 days = 9240 g of CO_2. Dividing this mass by the molecular weight of CO_2, it is calculated that the astronaut will produce a total of

$$\frac{9240\ g}{44.0\ g/mole} = 210 \text{ moles of } CO_2.$$

The number of moles of NaOH required to remove 210 moles

of CO_2 is 2×210 moles = 420 moles. Multiplying this by the molecular weight, the minimum amount of NaOH that must be carried on board is 420 moles × 40 g/mole = 16,800 g = 16.8 kg.

Q. 15 : Paper pulp (almost 100 % cellulose) is prepared commercially by digesting wood chips in a hot aqueous solution of calcium bisulfite, $Ca(HSO_3)_2$. The latter dissolves lignin and resins in the wood, leaving nearly pure cellulose. The sulfite solution is preapred by the folle ving reactions:

$$S + O_2 \rightarrow SO_2$$
$$SO_2 + H_2O \rightarrow H_2SO_3$$
$$CaCO_3 + 2H_2SO_3 \rightarrow Ca(HSO_3)_2 + CO_2 + H_2O$$

(a) For every 250 kg of limestone ($CaCO_3$) used in the process, what weight of sulfur would be required ? (b) What weight of $Ca(HSO_3)_2$ is produced from the limestone and sulfure in part (a) ?

Solution : From the stocihiometry of the above equations, 1 mole of sulfur yields 1 mole of H_2SO_3 and 1 mole $CaCO_3$ uses 2 moles of sulfur. The molecular weights of $CaCO_3$ and S are 100 g/mole and 32 g/mole, respectively. If 250 kg = 250,000 g of $CaCO_3$ are used then, the number of moles of $CaCO_3$ is

$$\frac{25,000 \text{ g } CaCO_3}{100 \text{ g / mole}} = 2500 \text{ moles.}$$

For every 2500 moles $CaCO_3$ used in the process, twice that amount, or 5000 moles, of sulfur is needed. The weight of sulfur required is (5000 moles) (32 g/mole) = 160,000 g = 160 kg.

For every mole of $CaCO_3$ consumed, one mole of $Ca(HSO_3)_2$ is produced. Thus, if 2500 moles of $CaCO_3$ is consumed then, 2500 moles of $Ca(HSO_3)_2$ are produced. To obtain the weight of $Ca(HSO_3)_2$ produced multiply its molecular weight (MW of $Ca(HSO_3)_2$ = 202 g/mole) by the number of moles produced. The weight of $Ca(HSO_3)_2$ produced is

$$(2500 \text{ moles}) (202 \text{ g/mole}) = 505,000 \text{ g} = 505 \text{ kg.}$$

Q. 16 : Lithopone white paint is made by heating a mixute of BaS and $ZnSO_4$ according to the equation

$$BaS + ZnSO_4 \rightarrow BaSO_4 + ZnS$$

What is the weight ratio of products $BaSO_4$ and ZnS in the paint ?

Solution : From the stoichiometry of this equation, 1 mole of BaS + 1 mole of $ZnSO_4$ → 1 mole of $BaSO_4$ → 1 mole of ZnS. Assuming that 1 mole of each product is contained in this paint, their weights would be equivalent to their molecular weights. The molecular weight of $BaSO_4$ is 233 g/mole; for ZnS it is 97 g/mole. The weight ratio is the ratio of weight of one substance compared to the weight of another substance. Thus, the weight ratio of $BaSO_4$ to ZnS in lithopone is

$$BaSO_4 : ZnS = 233 : 97 \quad \text{or} \quad 2.4 : 1$$

The initial assumption that only 1 mole of each product is contained in the paint is therefore not necessary. The weight ratio would be the same no matter how many moles of products were formed, since the products are formed in equimolar amounts.

Q. 17 : Manufacturers use soda-lime glass for containers. This type of glass can be prepared by melting sodium carbonate (Na_2CO_3), limestone ($CaCO_3$), and sand (SiO_2) together. The equation for this reaction is

$$Na_2CO_3 + CaCO_3 + 6SiO_2 \rightarrow \underset{\text{soda-lime glass}}{Na_2O.CaO.6SiO_2} + CO_2$$

If a manufacturer wanted to make 5000 wine bottles, each weighing 400 g, how many grams of sand would be requried ?

Solution : From the reaction equation, it can be seen that 6 moles of SiO_2 are consumed for every moles of soda–lime glass formed. Thus, one can determine the amount of sand needed to form the 5000 bottles once the number of moles of glass that makes up these bottles is found. To solve :

(1) determine the weight of the bottles and the number moles of soda-lime glass that this weight is equivalent to. (2) Calculate the number of moles of sand needed and from this its weight.

Solving :

(1) weight of bottles = (400 g/bottle) (5000 bottles)

$$= 2.0 \times 10^6 \text{ g.}$$

The molecular weight of soda-lime glass is 478 g/mole.

$$\text{number of moles} = \frac{2.0\times10^6 \text{ g}}{478 \text{ g / mole}} = 4.18\times10^3 \text{ moles of glass.}$$

(2) For each mole of glass formed, 6 moles of sand must react. Thus,

no. of moles of sand = $6 \times 4.18 \times 10^3 = 2.51 \times 10^4$ moles.

The molecular weight of sand is 60.1 g/mole. Hence, the amount of sand needed $= (2.51 \times 10^4 \text{ moles}) (60.1 \text{ g/mole})$
$= 1.51 \times 10^6$ g.

Q. 18 : An explorer, lost in the desert, stumbled on an abandoned automobile whose gas tank still held 5 liters of fuel. Gasoline has a density of 0.67 g/cc. Assuming gasoline to be composed of C_7H_{16} molecules, how many liters of water could he produce by burning the fuel and trapping the products ? The reaction is

$$C_7H_{16}\,(\ell) + O_2\,(g) \rightarrow CO_2\,(g) + H_2O\,(g)$$

Solution : There is one mole of H_2O produced by each mole of C_7H_{16} burned. To find the amount of water formed by burning the gasoline determine the number of moles of fuel present. Once the number of moles formed is determined, the number of grams of H_2O can be found.

There are 5 liters of gasoline present. It is given that 1 cc of C7H16 weighs 0.67 g. In 1 liter there are 1,000 cc, therefore the gasoline weights

$$5000 \text{ cc} \times 0.67 \text{ g/cc} = 3350 \text{ g}$$

One finds the number of moles present by dividing the total weight by the molecular weight. (MW of $C_7H_{16} = 100$).

$$\text{moles} = \frac{3350 \text{ g}}{100 \text{ g / mole}} = 33.50 \text{ moles.}$$

Because 33.50 moles of C_7H_{16} are present, 33.50 moles of H_2O can be formed. The number of grams of H_2O formed is found by multiplying the molecular weight of H_2O by the number of moles present (MW of $H_2O = 18$).

$$\text{grams} = 33.50 \text{ moles} \times 18.0 \text{ g/mole} = 603.0 \text{ g.}$$

Q. 19 : All the silicon and most of the oxygen in the earth's crust exists in the form of silicate minerals, such as granite. If the basic chemical unit of all silicates is the SiO_4^{4-} ion, which consists of an Si^{4+} ion and four O^{2-} ions, write the electronic configurations of the Si^{4+} and O^{2-} ion in silicate materials.

Solution : An electronic configuration of an element (or ion) gives a picture of the arrangement of electrons that surround a nucleus.

The atomic number gives the total number of electrons of the atom (not the ion). With this in mind, write the electron configuration of the oxygen and silicon atoms. The electronic configuration of oxygen with atomic number 8 is $1s^22s^22p^4$ and for silicon with atomic number 14 it is $1s^22s^22p^63s^2sp^2$.

To find the electronic configurations of the ions Si^{4+} and O^{2-} note each ion's charge. The charge of 4+ on Si indicates that it lost 4 electrons (electrons are negatively charged). Thus, if the silicon atom had a configuration of $1s^22s^22p^63s^23p^2$, and it loses 4 electrons, the configuration becomes $1s^22s^22p^6$ (ions will form by losing only high energy electrons).

The oxygen atom configuration was $1s^22s^22p^4$. The fact that the charge was 2– indicates that the atom gained two electrons. Consequently, the configuration becomes $1s^22s^22p^6$.

14

Pollution Problems

Q. 1 : One of the major atmospheric pollutants emitted by fuel-combustion power stations is a sulfur oxide mixture generally designated as So_x. It consists mainly of SO_2 but may contain anywhere from 1 to 10 % SO_3 computed as percent by weight of the total mass. What would be the approximate range to assign to x ?

Solution : One finds x by determining the number of O atoms present for each S present. For purposes of calculation assume that one always has 100 g of the mixture of SO_2 and SO_3. Thus when 1 % of the mixture is SO_3, there will be 1 g of SO_3 present and 99 g of SO_2 present. One then finds the number of moles present of each element. The ratio of S : O is then determined.

Solving for x when 1 % of the mixture is SO_3 : (MW of SO_2 = 64, MW of SO_3 = 80). The number of moles present is equal to the number of grams present divided by the molecular weight.

$$\text{For } SO_3: \frac{1\text{ g}}{90\text{ g / mole}} = 1.25\times10^{-2} \text{ moles}$$

$$\text{For } SO_2: \frac{99\text{ g}}{64\text{ g / mole}} = 1.55 \text{ moles.}$$

Because there is one S present in each compound the total amouunt of S is 1.55 moles + 1.25×10^{-2} moles or 1.56 moles. In 1 mole of SO_3 there are 3 moles of O, therefore in 1.25×10^{-2} moels of SO_3 there is three times that amount.

No. of moles of O in $SO_3 = 3 \times 1.25 \times 10^{-2} = 3.75 \times 10^{-2}$

There are 2 moles of O in 1 mole of SO_2, thus in 1.55 moles of

SO_2 there are 3.10 moles of O. The total number of moles of O present is equal to the sum of the moles contributed by the SO_2 and the SO_3.

Total No. of moles of O = 3.75 × 10^{-2} + 3.10 = 3.14 moles.

The formula for this compound is $S_{1.56}O_{3.14}$. The simplest formula is $SO_{2.01}$. When there is 1 % of SO_2 present, x = 2.01. x = 3.14/1.56.

When there is 10 % of SO_3 present, a similar method is used to find x.

(1) Find the number of moles present of SO_2 and SO_3.

(2) Determine the number of moles of S and O contributed by the SO_2 and SO_3.

(3) write the formula and simplify.

Solving for x :

(1) For a 100 g sample of gas where 10 % of the total weight is SO_3, 10 g of the gas is SO_3 and 90 g is SO_2.

$$\text{No. of moles of } SO_3 = \frac{10 \text{ g}}{80 \text{ g / mole}} = 1.25 \times 10^{-1} \text{ moles}$$

$$\text{No. of moles of } SO_2 = \frac{90 \text{ g}}{64 \text{ g / mole}} = 1.41 \text{ moles}$$

(2) The total number of moles of S present is equal to the sum of the number of moles of SO_2 and SO_3 present because each compound contributes 1 S per mole of compound present.

No. of moles of S = 1.25 × 10^{-1} + 14.1 = 1.54 moles

No. of moles of O from SO_3 = 3 × 1.25 × 10^{-1}

= 3.75 × 10^{-1} moles.

No. of moles of O from SO_2 = 2 × 1.41 = 2.82 moles.

Total No. of moles of O = 3.75 × 10^{-1} + 2.82 = 3.20 moles.

(3) The formula for this compound can be written $S_{1.54}O_{3.20}$ x= 3.20/1.54 = 20.8, thus the simplest formula is $SO_{2.08}$.

The range of x then is from 20.1 to 2.08.

Q. 2 : Analysis of the exhaust composition of the supersonic transport for one hour of flight yielded the following information :

Compound	Molecular Weight (g/mole)	Mass (g)
H_2O	18	8.0 × 10^7

CO_2	44	6.6×10^7
CO	28	3.6×10^6
NO	30	3.6×10^6

Determine the mole fraction of CO_2 in this mixture.

Solution : To determine the mole fraction of CO_2 in the mixture, we divide the number of moles of CO_2 by the total number of moles present. The number of moles of each component is found by dividing the mass of that component component by its molecular weight. Thus, for each component :

moles $H_2O = 8.0 \times 10^7$ g/18 g/mole $= 4.4 \times 10^6$ moles
moles $CO_2 = 6.6 \times 10^7$ g/44 g/mole $= 1.5 \times 10^6$ moles
moles $CO = 3.6 \times 10^6$ g/28 g/mole $= 1.3 \times 10^5$ moles
moles $NO = 3.6 \times 10^6$ g/30 g/mole $= 1.2 \times 10^5$ moles

The toal number of moles is

$4.4 \times 10^6 + 1.5 \times 10^6 + 1.3 \times 10^5 + 1.2 \times 10^5 = 6.2 \times 10^6$ moles.

The mole fraction of CO_2 is then

$$\text{mole fraction } CO_2 = \frac{\text{moles } CO_2}{\text{total moles}} = \frac{1.5\times10^6 \text{ moles}}{6.2\times10^6 \text{ moles}} = 0.24.$$

Q. 3 : Phosphate can be removed from sewage effluent by adding magnesium oxide and ammonium ion according to the reaction

$$5H_2O + H_2PO^-_4 + MgO + NH_4^+ \rightarrow Mg(NH_4)PO_4.6H_2O.$$

The magnesium ammonium phosphate hexahydrate (MG $(NH_4)PO_4.6H_2O$) precipitates out. In theory, how much of this could be produced if a city processes 5 million gal. (41.7 million 1b.) daily of wastewater containing 30 ppm $H_2PO^-_4$? (1 pound = 453.59 g).

Solution : The best way to approach this problem is to calculate how much $H_2PO^-_4$ is present in grams. Once the weight in grams is known one can calculate from the stoichiometry of the equation, the number of moles of Mg $(NH_4)PO_4.6H_2O$ that will precipitate.

The wastewater has 30 ppm (parts per million) of $H_2PO_4^-$. In other words, 30 lbs. of $H_2PO^-_4$ are present in one million 1bs. of of wastewater. The city produces 41.7 million 1bs. wastewater. Thus, 30 (41.7) = 1,251 1bs. of $H_2PO^-_4$ should be present. Because 1 pound = 453.59 grams; 1,251 1bs = 567,454 grams.

Compute how many moles of $H_2PO^-_4$ are present. A mole is defined as grams (weight)/molecular weight (M.W.). The M.W. of

$H_2PO^-_4$ = 96.97 g/mole. The number of moles of $H_2PO^-_4$ is

$$= \frac{567{,}454 \text{ g}}{96.97 \text{ g / mole}} = 5852 \text{ moles.}$$

According to the reaction equation, one mole of $H_2PO^-_4$ generates one mole of precipitate. The M.W. of the precipitate = 245 g/mole. Recalling the definition of a mole, the amount of Mg $(NH_4)PO_4.H_2O$ = (5852) (245) = 1.433×10^6 g = 1.433×10^3 kg.

Q. 4 : The SO_2 content in air may be determined by passing the air through water to produce a solution of sulfurous acid, H_2SO_3. The subsequent reaction of this solution with potassium permanganate, $KMnO_4$, establishes the amount of SO_2 absorbed from the air. (a) Write the balanced equation for the reaction of SO_2 with water to produce H_2SO_3. (b) Write the balanced equation for the reaction of H_2SO_3 (or SO^{2-}_3) with $KMnO_4$ (or MnO^-_4) to form $MnSO_4$, K^+ and H_2SO_4, among other products. (c) If 1000 liters of air are passed through water and the resulting solution requires 2.5×10^{-5} moles of $KMnO_4$ for complete reaction (part b), what is the concentration (in ppm) of SO_2 in the air. Assume STP conditions.

Solution : (1) An SO_2 mlecule reacts with a water molecule to form an H_2SO_3 molecule according to the following equation :

$$SO_2 + H_2O \rightarrow H_2SO_3$$

However, H_2SO_3 is a diprotic acid and dissociates into 2 protons and a sulfate ion

$$H_2SO_3 \rightarrow 2H^+ + SO_3 =$$

Thus, for every mole of SO_2 consumed 1 mole of SO_3 = are produced.

(b) This reaction is a typical redox reaction and the equation must be balanced using the standard rules.

From the problem, one knows initially that

$$H_2SO_3 + KMnO_4 \rightarrow MnSO_4 + H_2SO_4 + K^+$$

This reaction is not balanced in terms of charge and elemental constituents. The first rule is to separate the above reaction into a reduction and an oxidation reaction.

red: $KMnO_4 \rightarrow MnSO_4 + K^+$

Mn is reduced from +7 to +2 state.

ox: $H_2SO_3 \rightarrow H_2SO_4$

S is oxidized from +4 to +6 state.

Next one balances the oxygen and hydrogen atoms in these reactions using OH^-, H^+, and H_2O molecules on either side of the equation depending upon whether the reaction is acidic or basic. once the masses are balanced, electrons are used to blanace charges and are placed on the appropriate side of the equation. Thus,

red: $KMnO_4 \rightarrow MnSO_4 + K^+$

The potassium ion and the SO_4 ion do not play significant roles in the balancing of this equation and are therefore unnecessary.

$$MnO^-_4 \rightarrow Mn^{+2}$$

This reaction is acidic and therefore will be balanced using H^+ and H_2O.

$$8H^+ + MnO^-_4 \rightarrow Mn^{+2} + 4H_2O$$

The H_2O is placed on the right to balance the oxygen in MnO_4^-; the H in H_2O is balanced on the left with H^+. The equation will be balanced in terms elements and charge when electrons are added. Thus,

$$8H^+ + 5e^- + MnO_4^- \rightarrow Mn^{++} + 4H_2O$$

ox: $H_2SO_3 \rightarrow H_2SO_4$

the hydrogen ions are unnecessary and play insignificant roles in balancing this equation; therefore,

$$SO_3^= \rightarrow SO_4^=$$

Next place water on the left and balance using a sufficient number of hydrogen ions on the right.

$$H_2O + SO_3^= \rightarrow SO_4^= + 2H^+$$

Now balance off the charge using electrons:

$$H_2O + SO_3^= \rightarrow SO_4^= + 2H^+ + 2e^-$$

Once both oxidation and reduction processes are balanced, they are added together in the following manner.

red: $[8H^+ + 5e^- + MnO^-_4 \rightarrow Mn^{++} + 4H_2O] \times 2$

ox: $[H_2O + SO_3^= \rightarrow SO_4^= + 2H^+ + 2e^-] \times 5$

These reactions are solved simultaneously to eliminate the electrons in both reactions. Thus,

$$16H^+ + 2MnO^-_4 \rightarrow 2Mn^{++} + 8H_2O$$

$$5H_2O + 5SO_3^= \rightarrow 5SO_4^= + 10H^+$$

Total: $6H^+ + 2MnO^-_4 + 5SO_3^= \rightarrow 2Mn^{++} + 5SO_4^= + 3H_2O$

(c) To solve this part of the problem, one must know the number of moles of SO_2 in air is equal to the number of moles of SO_3 = that are represented in the equation for the reaction with KMnO4. This amount is determined from the stoichiometry.

For every mole of SO_2 consumed, 1 mole of $SO_3^=$ is formed, and for every 1 mole of $SO_3^=$ consumed, 2/5 mole of MnO^-_4 is also consumed. Thus,

number moles of SO_2 = 5/2 (2.5×10^{-5}) = 6.25×10^{-5} moles

To obtain a concentration in units of ppm (parts per million), one needs to know the volume of SO_2 consumed and the total volume of air. The total volume of air is 1×10^3 liters. The volume of SO_2 consumed is

$$\text{volume of } SO_2 = (6.25 \times 10^{-5} \text{ moles})(22.4 \text{ liter/mole})$$
$$= 1.4 \times 10^{-3}\ \ell,$$

since, at STP, one mole of gas will occupy 22.4 liters.

Thus, the ratio of SO_2 volume to total volume of air is

$$\frac{1.4\times10^{-3} \text{ liter}}{1\times10^{3} \text{ liter}} = 1.4\times10^{-6}.$$

To convert this ratio into ppm, we make the denominator 10^6 liters,

$$\frac{1.4\times10^{-3} \text{ liter}}{1\times10^{3} \text{ liter}} \times \frac{10^3}{10^3} = \frac{1.4}{1\times10^{6}}$$

The concentration of SO_2 in the air is 1.4 ppm.

Q. 5 : An approach to controlling industrail pollution is to recover waste products in usable form. In the reaction,

$$2H_2S + SO_2 \rightarrow 3S + 2H_2O$$

waste H_2S and SO_2 are recovered as elemental sulfur liquid. How much sulfur can be recovered if 650,000 kg of SO_2 are deposited into the air ?

Solution : From the stoichiometry of this equation, 2 moles of H_2S react with 1 mole of SO_2 to produce 3 moles of S and 2 moles of H_2O.

To solve this problem, calculate how many moles of SO_2 are polluted into the air by dividing its weight by its molecular weight (MW SO_2 = 64 g/mole).

From the stoichiometry of the equation, determine how many moles of S would be produced, and then calculate the weight of S by multiplying the number of moles of S times its molecular weight (MW S = 32 g/mole). Thus,

$$\text{number moles of } SO_2 = \frac{65{,}000{,}000 \text{ g } SO_2}{64 \text{ g / mole}} = 1.02 \times 10^7 \text{ moles}$$

From the stoichiometry of the equation 1 mole $SO_2 \rightarrow 3$ mole S or 1.02×10^7 mole $SO_2 \rightarrow 3.06 \times 10^7$ mole S.

weight of S = (3.06×10^7 mole S) (32 g/mole)

= 97.9×10^7 g = 979,000 kg.

Q. 6 : Iodine pentoxide (I_2S_5) is a very important reagent because it can oxidize carbon monoxide (CO), which is a major pollutant of the atmosphere, to carbon dioxide (CO_2) in the absence of water. The reaction involved is

$$I_2O_5 + 5CO \rightarrow I_2 + 5CO_2$$

How many kg of I_2O_5 is needed to completely 140,000 kg of CO from the air ?

Solution : From the stoichiometry of this equation, 1 mole of I_2O_5 completely reacts with 5 moles of CO. To solve this problem, calculate the number of moles of CO that will react by dividing its weight by its molecular weight of 28 g/mole. Determine, also, how many moles of I_2O_5 will completely react with this, and then calculate the weight of I_2O_5 that will completely react by multiplying the number of moles of I_2O_5 by its molecular weight (MW of I_2O_5 = 334 g/mole). Thus,

$$\text{number moles CO} = \frac{14{,}000{,}00 \text{ g CO}}{28 \text{ g / mole}} = 5.00 \times 10^6 \text{ mole.}$$

This means that one fifth of this number of moles will be the moles of I_2O_5 required. Recall, there will exist 5 moles of CO for every mole of I_2O_5. Therefore,

$$\text{moles } I_2O_5 = \frac{1}{5}(5 \times 10^6) = 1 \times 10^5 \text{ moles.}$$

This means that

(1×10^6 moles) (334 g/mole) = 3.34×10^8 grams of I_2O_5 is needed to completely clear 140,000 kg of CO from the air.

Q. 7 : The average concentration of particulate lead in a city with highly polluted air is 5 μg/m³. If an adult respires 8500 liters of air daily and if 50 per cent of the particles below 1 μm in size are retained into the lungs, calculate how much lead is abosrbed in the lungs each day. Of the particulate lead in urban air, 75 per cent is less than 1 μm in size.

Solution : To solve this problem (1) determine the total percentage of particulate lead reamining in the lungs, (2) determine the amount of lead absorbed into the lungs.

The problem states 75 % of particulate lead is less than 1 μm in size, therefore, 25 % of the lead is greater than 1 μm in size and does not reamin inside the lungs. Of the 75 % particulate lead less than 1 μm in size, only 50 % of this is retained in the lungs. The total percentage of lead retained is 37.5 %.

The total volume of air-intake is 8500 liters, so that total volume of air-intake = (8500) (10^{-3} m³/liter) = 8.5 m³

The total amount of lead-intake is

amount of lead = (8.5 m³) (5 μg/m³) = 42.5 μg lead.

The total amount of lead absorbed is

amount of lead absorbed = (42.5 μg) (0.375) = 15.9 μg.

Q. 8 : In 1932, many cities showed higher sunlight levels than in the previous year. Explain this observation.

Solution : To answer this problem, consider the effect of atmospheric particles on the amount of sunlight received and what happened in 1932.

By 1932 three years of slackened industrial activity had passed. Then as now, industries released massive amounts of particulate matter, i.e., solid and liquid aerosols, into the atmosphere. These particles have the ability to scatter and absorb light. Because industrial activity was drastically reduced, fewer particles were released that could scatter or absorb the sunlight. The citites, therefore, received more sunlight in 1932 than in years when industry was thriving.

It is interesting to note that because of particulate matter, cities receive about 15 to 20 % less solar radiation than do rural areas.

Q. 9 : The air pollutant, sulfur dioxide, is considered particularly foul. Explain why there is so much concern with this compound.

Solution : To solve this problem, consider the reactions that can occur between sulfur dioxide (SO_2) and other elements, and/or compounds, in the natural envrionment.

SO_2 can enter the atmosphere from both natural and man-made sources. Besides causing eye irritation and plant damage, it can cause other health problems because of its ability to form acids in the atmosphere. When SO_2 comes in contact with water (H_2O), the following reaction occurs.

$$SO_2 + H_2O \rightarrow \underset{\text{sulfurous acid}}{H_2SO_3}$$

SO_2 can also be oxidized in air to sulfur trioxide (SO_3) by the reaction

$$2SO_2 + O_2 \rightarrow 2SO_3.$$

Sulfur trioxide, when exposed to water, produces sulfuric acid by the reaction

$$SO_3 + H_2O \rightarrow \underset{\text{sulfuric acid}}{H_2SO_4}$$

These two acids, sulfuric and sulfurous, account for the corrosion of metal parts exposed to the atmosphere. They are also responsible for the decay of stone statues and buildings. This occurs because most stones consist of $CaCO_3$ (calcium carbonate), which, upon addition of H_2SO_4, undergoes the following reaction :

$$CaCO_3 + H_2SO_4 \rightarrow CaSO_4 + H_2O + CO_2.$$

$CaSO_3$ is soluble in water and will wash away with the rain causing the wearing away of the stone.

Q. 10 : If the carbon dioxide content of the air doubled, what

$$\text{mass} = \frac{(1.013\times10^5 \text{ kg} - \text{m}/\text{m}^2\text{s}^2)(4\pi)(6.37\times10^6 \text{m})^2}{(9.8 \text{ m}/\text{sec}^2)}$$

$$= 5.27\times10^{18} \text{ kg.}$$

Q. 11 : Explain what is meant by primary, secondary, and tertiary treatment of sewage.

Solution : Sewage treatment is carried out in three stages known as primary, secondary, and tertiary. Each uses a different process to eliminate different substances from the sewage.

Primary treatment uses mechanical means to separate out solid objects such as sticks and rags. These solid objects are removed by means of a coarse screen. The liquid that passes through this screen flows into settling tanks, where insolubvle material settles and forms sludge. This removes approximately one third of the pollutants.

Whereas primary treatment is mechanical, secondary treatment is biological. the liquid or effluent from the primary facility filters slowly through a bed of rock. Bacteria present consume most of the organic material. In many plants, oxygen is bubbled in to accelerate the growth of bacteria and, as such, the decomposition of the organic waste. This process removes about 90 % of the biodegradable materials in the water.

The last type of treatment, tertiary, is the most expensive. It is used to remove organic chemical, nutrietns, and excessive salts. To accomplish this, the treatment may include such processes as chemical coagulation, distillation and reverse osmosis.

Q. 12 : What is eutrophication ? What problems are associated with it ? Why would a reduction of phosphate ions from waterways alleviate eutrophication in lakes and streams ?

Solution : The enrichment of water with nutrients is a naturally occurring biological process called eutrophication. It is caused by excess nutrients in water that enable algae to grow to great abundance. The problem of massive algae growth occurs when it dies. As it dies, it sinks to the bottom of lakes and rivers and begins to decay. This decaying process consumes oxygen thereby deleting the oxygen content of the water. Fish and other aquatic life dependent on oxygen suffocate. Therefore, the decaying of excessive amounts of algae can lead to the death of marine life. This is one major problem associated with eutrophication.

A second problem also results from the depletion of oxygen. Remember that anaerobic bacteria, i.e. bacteria that can live without oxygen, still function effectively under the conditions described above. These bacteria feed on decaying algae. As a result of their metabolic process, they produce compounds such as hydrogen sulfide, which cause the putrid odor associated with decaying organic matter. The water becomes foul smelling and and foul tasting.

A reduction of phosphate would alleviate eutrophication because it is one of the major nutrients of algae. A smaller amount of phosphate in the water would limit the growth of algae.

Q. 13 : The combined effects of pollution have been known to bring about the phenomena of synergism and antagonism. Explain each of these terms. Provide specific exmples of each.

Solution : The effects of individual pollutants can be **disastrous** to health, while the combined action of two or more pollutants can be completely different on health and/or environment.

When the combined effects of two or more **polluntants** are more severe or qualitatively different from the individual effects, a condition called synergism exists. For example, in 1966 and 1967, there was injury to peanut crops due to the synergistic action of ozone and sulfur dioxide.

A condition termed antagonism exists when the combined effects are less severe. For example, cyanides in industrail wastes are quite poisonous to aquatic life, however, in the presence of nickel, a nickel-cyanide complex is formed. This nickel-cyanide complex is not as toxic as the cyanides originally produced.

Q. 14 : (A) Explain the purpose of a coliform bacteria count in natural water. (B) What is meant by hard or refractory organic compounds ?

Solution : (a) One of the most important problems associated with polluted water is the possible presence of disease-causing bacteria. A test for the presence of bacteria is made by measuring the amount of common coliform bacteria in water. (Coliform bacteria are harmless and reside in the large intentines of human). Coliform bacteria are present in plentiful quantities in feces, which become part of sewage wastes. Since coliform bacteria do not thrive in an aquatic environment, they give an indication of how recently and to what extent sewage pollution has occurred by their count (which is proportioanl to the amount of disease-causing bacteria present). If the count is high, human sewage contamination has taken place, possibly accompanied by the presence of dangerous bacteria.

(b) Hard or refractory organic compounds are nondegradable substances that decay very slowly in water. Degradation refers to the breaking down of substances into simpler forms. Examples of nondegradable substances are certain pesticides and detergents.

Q. 15 : A chemist has a quantity of steam. He wants to determine whether or not it has been polluted with acidic mine drainage (AMD). How should he go about determining this ?

Solution : AMD occurs yearly when millions of tons of sulfuric acid seep out from coal and copper mines to pollute waterways.

It begins by the oxidation of iron pyrite (FeS_2) to sulfate :

$$FeS_2 \xrightarrow[H_2O]{O_2} Fe^{2+} + So_4^{2-}$$

The ferrous ions are then oxidized to ferric ions :

$$Fe^{2+} \xrightarrow[H_2O]{O_2} Fe^{3+}$$

This process can be sped up by microorganisms. The ferric ions can react with water to give hydrogen ions and a precipitate of hydrated ferric oxide :

$$Fe^{3+} \xrightarrow{H_2O} Fe_2O_3.H_2O + H^+$$

Thus, acidic mine drainage is a diluted sulfuric acid solution carrying iron. If the steam has AMD, it must carry iron. Addition of thiocyanate to ferric (iron) solutions will yield an intensely red-colored complex ion. The reaction is as follows :

$$\underset{}{Fe^{3+}} + \underset{\text{thio-cyanate}}{SCN^-} \rightarrow \underset{\text{red-colored complex ion}}{FeSCN^{2+}}$$

Thus, the chemist can detect for AMD pollution in the steam by adding thiocyanate. If a red-colored material is formed, he knows it is contaminated.

Q. 16 : A chemist dilutes two 10–ml samples of waste water to 300 ml with aerated water. One sample is analyzed for dissolved oxygen at 20°C immediately, but the second sample is incubated for 5 days before being analyzed for dissolved oxygen. The results are

$$(DO)_{0\text{ days}}^{20°C} = 7.9 \text{ mg / liter and } (DO)_{5\text{ days}}^{20°C} = 1.0 \text{ mg / liter}$$

What is BOD_5 of the waste water ?

Solution: BOD stands for Biochemical Oxygen Demand. It measures the potential oxygen requirement of sewage and of sewage polluted waters. $(DO)^{20°C}$ stands for the concentration of dissolved oxygen at 20°C.

The numerical value of BOD can be found by measuring the dissolved oxygen content (DO) of the sample before and after the 5 day incubation period. BOD is the difference between these dissolved oxygen concentrations (after accounting for corrections of dilution and any additional organic material that might be present in the diluting water).

Proceed as follows :

The 10 ml sample was diluted to 300 ml or 0.3 ℓ. The DO concentrations at day 0 and day 5 are multiplied by 0.3 ℓ to determine the amount of oxygen used on those days

$(7.9 \text{ mg}/\ell)(0.3\ \ell) = 2.37$ mg Day 0

$(1.0 \text{ mg}/\ell)(0.3\ \ell) = 0.3$ mg Day 5

These amounts are subtracted to determine the BOD of the sample.

$$2.37 - 0.3 = 2.07 \text{ mg}$$

The original sample was 10 ml; the density of water is 1 g/ml. Thus, 10 ml weights 10 g. The BOD of this sample is 2.07 mg/10 g. BOD's are usually expressed in ppm. Converting 2.07 mg/10 g to ppm one ontains 207 ppm.

For a diversified warm-water biota, including game fish, the DO concentrations should be at least 5 mg/ℓ or 5 ppm. A body of water is considered polluted when the DO concentration drops below the level necessary to sustain a normal biota for that water. The BOD range for untreated municipal sewage is 100–400 ppm.

Q. 17 : During the course of World War I, 1.1×10^8 kg of poison gas was fired on Allied soldiers by German troops. If the gas is assumed to be phosgene ($CoC1_2$), how many molecules of gas does this correspond to ?

Solution: To solve this problem we must convert mass to number of moles and then multiply by Avogadro's number to obtain the corresponding number of molecules.

The number of moles of gas is given by

$$\text{moles} = \frac{\text{mass (grams)}}{\text{molecular weight of } COCl_2 \text{ (g / mole)}}$$

The mass of gas is 1.1×10^8 kg $= 1.1 \times 10^8$ kg $\times 10^3$ g/kg $= 1.1 \times 10^{11}$ g. The molecular weight of $COC1_2$ is obtained by adding the atomic weights (atm. wgt.) of its constituents. Thus,

$$\begin{aligned} \text{molecular weight } (COC1_2) &= \text{atm wgt (C)} + \text{atm wgt (0)} \\ &\quad + 2 \times \text{atm wgt (C1)} \\ &= 12.0 \text{ g/mole} + 16.0 \text{ g/mole} \end{aligned}$$

$$+ 2 \times 35.5 \text{ g/mole}$$
$$= 99 \text{ g/mole.}$$

Hence, the number of moles of gas is

$$\text{moles} = \frac{\text{mass}}{\text{molecular weight}} = \frac{1.1 \times 10^{11}\text{g}}{99 \text{ g.mole}}$$

$$\cong 1.1 \times 10^{9} \text{ moles.}$$

Multiplying the number of moles by Avogadro's number, we obtain the number of molecules of gas :

$$\text{number of molecules} = \text{moles} \times \text{Avogadro's number}$$
$$= 1.1 \times 10^{9} \text{ moles} \times 6 \times 10^{23} \text{ molecules/mole}$$
$$= 6.6 \times 10^{32} \text{ molecules.}$$

Q. 18 : For a single year, the motor vehicles in a large city produced a total of 9.1×10^{6} kg of the poisonous gas carbon monoxide (CO). How many moles of CO does this correspond to ?

Solution: The number of moles of a substance is equal to the quotient of the mass (in grams) of that substance and its molecular weight (in g/mole), or

$$\text{moles} = \frac{\text{mass (g)}}{\text{molecular weight (g / mole)}}$$

The mass of CO is 9.1×10^{6} kg $= 9.1 \times 10^{6}$ kg $\times$ 1000 g/kg. = 9.1×10^{9} g. The molecular weight of CO is the sum of the atomic weight of C and the atomic weight of O, or molecular weight of (CO) = atomic weight (C) + atomic weight (O)

$$= 12 \text{ g/mole} + 16 \text{ g/mole}$$
$$= 28 \text{ g/mole.}$$

Hence,

$$\text{moles} = \frac{\text{mass (g)}}{\text{molecular weight (g / mole)}} = \frac{9.1\times10^{9}\text{g}}{28 \text{ g / mole}}$$

$$= 3.3\times10^{8} \text{ moles.}$$

Q. 19 : An automobile travelling at 10 miles per hour produces 0.33 lb of CO gas per mile. How many moles of CO are produced per mile ?

Solution: The number of moles of a substance is equal to the quotient of the mass (in grams) of that substance and its molecular weight (in g/mole), or

$$\text{moles} = \frac{\text{mass (g)}}{\text{molecular weight (g / mole)}}$$

The mass of CO is 0.33 lb = 0.33 lb × 454 g/lb = 150 g. The molecular weight of CO is the sum of the atomic weight of C and the atomic weight of O, or

$$\begin{aligned}\text{molecular weight (CO)} &= \text{atomic weight (C)} + \\ &\quad \text{atomic weight (O)} \\ &= 12 \text{ g/mole} + 16 \text{ g/mole} \\ &= 22 \text{ g/mole.}\end{aligned}$$

Hence,

$$\text{moles} = \frac{\text{mass (g)}}{\text{molecular weight (g / mole)}} = \frac{150\text{g}}{28 \text{ g / mole}}$$

$$= 5.4 \text{ moles per mile.}$$

Q. 20 : The maximum safe concentration of fluoride ion in drinking water is 2.0 ppm (parts per million). Express this concentration in milligram%.

Solution: Milligram percent is the number of milligrams of solute in 100 ml. of liquid. It is given that there are 2 g. of fluoride in 1.0×10^6 g H_2O. Because 1 ml of H_2O weights 1 g, there are 2 g of fluoride in 1.0×10^6 ml of H_2O. Solving for mg%:

$$\text{milligram\%} = \frac{100 \text{ ml} \times 2\text{g} \times 1000 \text{ mg / g}}{1 \times 10^6 \text{ ml}} = .20 \text{ mg\%.}$$

The concentrations of substances in body fluids such as urine or blood are rather low, and for this reason they are generally expressed in milligram%.

Q. 21 : Suppose a rain cloud has a water volume of 10^5 liters. If the cloud forms over a chemical factory which puts SO_3 into the air and each SO_3 molecule bonds with one out of 10^4 H_2O molecules (to form H_2SO_4), what is the molarity of the H_2SO_4 in the cloud ?

Solution: To solve this problem, the number of moles of H_2SO_4 must be determined. This is done by applying the stoichiometric rules to the equation of this particular reaction :

$$H_2O + SO_3 \rightarrow H_2SO_4$$

According to this equation for every mole of SO_3 consumed, one mole of H_2SO_4 is produced or, in other words, if the number of moles of SO_3 consumed is known, then the number of moles of H_2SO_4 produced is also known. The number of moles of SO_3 can be found since it is the same number as H_2O (from the stoichiometry of the equation). The number of moles of H_2O consumed can be found by dividing its weight by its molecular weight (MW = 18.02) or

$$\text{number of moles of } H_2O = \frac{(10^5 \text{ liter})(1000 \text{ g/liter})}{18.02 \text{ g/mole}}$$

$$= 5.5 \times 10^6 \text{ moles}$$

Note: 1000 g/liter is the density of water (rain).

However, the number of water molecules is needed and thus,

number of H_2O molecules

$$= (5.5 \times 10^6 \text{ moles})(6.02 \times 10^{23} \text{ molecules/mole})$$

$$= 3.3 \times 10^{30} \text{ molecules.}$$

The problem states that SO_3 molecules bond with one out of 10^4 H_2O molecules. From this, the number of SO_3 molecules is determined,

$$\frac{3.3 \times 10^{30} \text{ } H_2O \text{ molecules}}{10^4 \dfrac{H_2O \text{ molecules}}{SO_3 \text{ molecules}}} = 3.3 \times 10^{26} \text{ } SO_3 \text{ molecules.}$$

The number of moles of SO_3 can now be obtained.

$$\text{number of moles of } SO_3 = \frac{(3.3 \times 10^{26} SO_3 \text{ molecules})}{6.02 \times 10^{23} \dfrac{\text{molecules}}{\text{mole}}}$$

$$= 5.5 \times 10^2 \text{ moles}$$

or 5.5×10^2 moles of H_2SO_4 are formed.

The molarity of the H_2SO_4 produced is thus

$$\frac{5.5 \times 10^2 \text{ moles}}{10^5 \text{ liter}} = 5.5 \times 10^{-3} \text{ M.}$$

□□□